REVISE BTEC

Application of Science UNIT 8 REVISION WORKBOOK

Series Consultant: Harry Smith

Author: Jennifer Stafford-Brown

THE REVISE BTEC SERIES

Application of Science Revision Workbook	9781446902844
Application of Science Revision Guide	9781446902837
Principles of Applied Science Revision Workbook	9781446902783
Principles of Applied Science Revision Guide	9781446902776

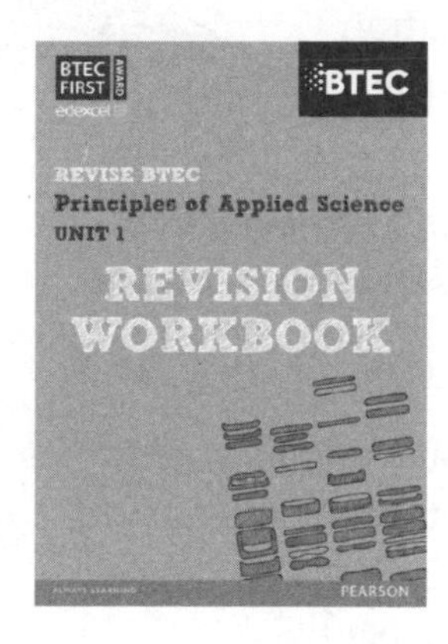

This Revision Workbook is designed to complement your classroom and home learning, and to help prepare you for the external test. It does not include all the content and skills needed for the complete course. It is designed to work in combination with Edexcel's main BTEC Applied Science 2012 Series.

To find out more visit:
www.pearsonschools.co.uk/BTECsciencerevision

1-to-1 page match with the Revision Guide ISBN 9781446902837

Contents

A small bit of small print

Edexcel publishes Sample Assessment Material and the Specification on its website. This is the official content and this book should be used in conjunction with it. The questions in this book have been written to help you practise every topic in the book. Remember: the real exam questions may not look like this.

You will need a calculator for your Unit 8 Scientific Skills paper.

You may find it helpful to practice using one with the questions in this book.

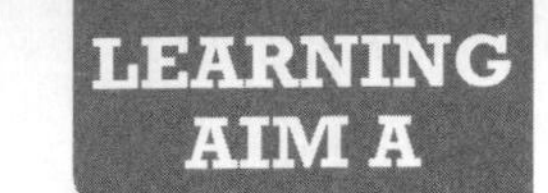

Scientific equipment 1

1 Which **one** of the following pieces of scientific equipment would be most suitable for mixing substances? Put a cross in the box next to the correct answer. **(1 mark)**

☐ **A** test tube

☐ **B** beaker

☐ **C** conical flask

☐ **D** measuring cylinder.

It needs to be large enough for a stirring rod to fit in and stir the substances.

2 Explain why a stirring rod should be clean before it is used to mix a substance. **(2 marks)**

....

....

3 Identify the following pieces of equipment. **(3 marks)**

A ....

Choose from the following:
Bunsen burner
heatproof mat
rubber tubing.

C ....

B ....

Guided

4 List **three** pieces of scientific equipment that are used to hold materials being heated. **(3 marks)**

A beaker can be heated, a ....

....

....

Scientific equipment 2

1 Which piece of equipment would be best for heating a small amount of liquid in a Bunsen flame? Put a cross in the box next to the correct answer. **(1 mark)**

☐ **A** test tube

☐ **B** boiling tube

☐ **C** beaker

☐ **D** conical flask.

There is only a small amount of liquid that needs to be heated. The equipment must be able to withstand high temperatures as it is going to be put in the Bunsen burner flame.

2 Max heats a test tube containing water over a Bunsen burner. He uses tongs to hold the test tube.

Guided

(a) Explain why tongs are used to hold the test tube over a Bunsen burner flame. **(2 marks)**

The Bunsen burner will heat up the test tube and ....

....

The tongs are used so that ....

....

(b) Describe other pieces of scientific equipment that Max could use instead of tongs to hold the test tube over the Bunsen burner flame. **(2 marks)**

....

(c) Describe why test tube racks are used to store test tubes. **(2 marks)**

....

(d) Describe a piece of scientific equipment that could be used to mix a solid into the water, e.g. sugar. **(1 mark)**

....

Chemistry equipment

1 Identify a piece of laboratory equipment that could be used to accurately measure 5 cm^3 of glucose solution for an experiment. **(1 mark)**

The word 'measure' is important here.

....

2 Sian is carrying out a neutralisation reaction. In her experiment, she adds sodium hydroxide to nitric acid. She needs to be able to measure the pH of the solution so that she can see when the neutralisation has taken place.

Name **two** different pieces of scientific equipment that Sian could use to test the pH of the solution. **(2 marks)**

....

....

3 Ian wants to separate a mixture of water and sand. He chooses a filter and filter funnel as shown in the diagram below.

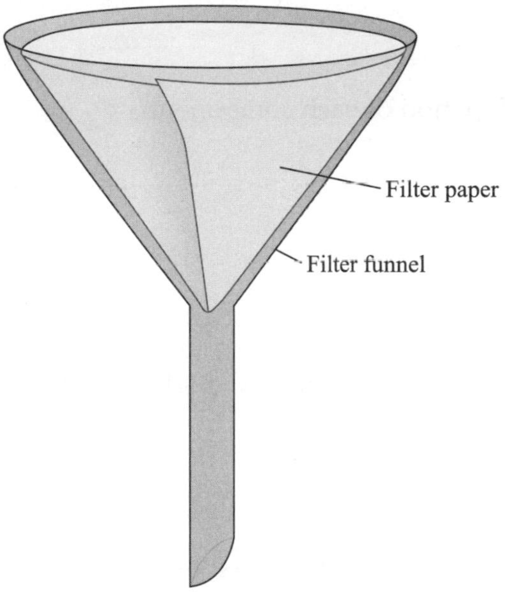

Filter paper

Filter funnel

Guided

(a) Explain how this equipment is used for this experiment. **(3 marks)**

Filter paper and a filter funnel are used to ....

....

The mixture is then poured onto the filter paper and the water ....

....

(b) Name a piece of scientific equipment that could be used to find the mass of the sand that is collected in the filter paper. **(1 mark)**

....

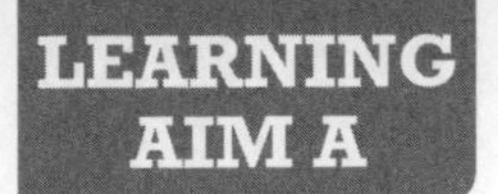

Physics equipment

1 Which of the following pieces of equipment could be used to measure the force of an object? Put a cross in the box next to the correct answer. **(1 mark)**

☐ **A** power pack ☐ **B** voltmeter

☐ **C** Newton meter ☐ **D** multimeter.

What is the unit of force?

2 Damien sets up an electric circuit. A diagram of the circuit is shown here.

(a) Identify the components in the circuit. **(3 marks)**

A

....

....

V

....

(b) Describe the function of each component. **(3 marks)**

....

....

....

(c) Identify a component that Damien could add to the circuit to reduce the current. **(1 mark)**

....

Guided **3** Earl is an athletics coach. He is carrying out an investigation into the acceleration of his athletes over a 50 m distance. In his investigation he plans to measure the times that the athletes take to run to 10, 20, 30, 40 and 50 m.

Describe the scientific equipment that he could use to measure the times at these intervals in the 50 m run. **(2 marks)**

He could use light gates set up at each of the distance intervals that he wants to measure.

These work by recording the ....

....

....

Write one more sentence about how light gates work.

Biology equipment

1 Susan is a forensic scientist. She uses a microscope to look for traces of skin and hair that can be used as evidence of a crime. A microscope can be used to view things that are too small to be seen by the naked eye. The magnification of a microscope refers to how much the object being observed is enlarged.

(a) Put a cross in the box next to the correct answer.

If the microscope lens used is 10×, this means the cell being viewed is: **(1 mark)**

☐ **A** 100 times as big

☐ **B** 1000 times as big

☐ **C** 10 times as big

☐ **D** 0.1 times as big.

(b) Describe the scientific equipment that Susan will need in order to prepare a sample to be looked at through a microscope. **(2 marks)**

....

....

....

There are two marks for this question so you will need to describe two pieces of equipment.

Guided

2 Ahmed is carrying out an investigation into the effect of different doses of penicillin on a type of bacteria. He needs to grow five colonies of bacteria, and then he will add different doses of penicillin to each to work out which dose is the most effective.

Describe the scientific equipment Ahmed will need to grow a colony of bacteria. **(3 marks)**

He will need a strain of bacteria, ...

....

....

3 A health visitor is investigating the effect of exercise on children aged between 8 and 9. For this investigation, she will need to find out the height and mass of each child.

Describe in detail how the health visitor could find out the height and mass of these children. **(4 marks)**

....

....

....

....

Had a go ☐ Nearly there ☐ Nailed it! ☐

Risks and management of risks

1 Which of these is an example of a risk? Put a cross in the box next to the correct answer. **(1 mark)**

☐ **A** a Bunsen burner

☐ **B** burning your skin

☐ **C** a pair of goggles

☐ **D** a toxic chemical.

A risk is the harm that could be caused and the chances of it happening.

Guided **2** Identify **two** hazards and **two** risks in this image. **(4 marks)**

One hazard is the Bunsen burner. ....

One risk is her loose hair. ....

3 Safety goggles are worn during many different scientific experiments to protect a person's eyes – this is known as a control measure.

What is a control measure? **(2 marks)**

....

....

4 Paul is carrying out an experiment in which two chemicals are mixed together and chlorine gas is produced. Chlorine gas is toxic.

Describe and explain the control measures that should be used to carry out this experiment. **(2 marks)**

....

....

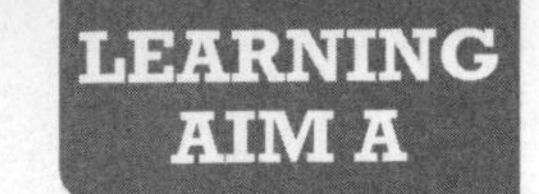

Hazardous substances and control measures

1 Put a cross in the box next to the hazard symbol for a substance that is corrosive. **(1 mark)**

Corrosive means it dissolves skin and other materials.

☐ A ☐ B ☐ C ☐ D

2 Val is a science technician. She must fully understand the hazard signs on the chemicals and substances that she uses so that she knows how they should be stored and handled safely.

(a) Magnesium has the hazard symbol . State what the hazard symbol means and describe how this substance should be stored. **(2 marks)**

....

....

Guided

(b) A strong acid has the hazard symbol . Describe the control measures she must take with this type of chemical. **(2 marks)**

It is corrosive so she must ....

....

(c) Sodium hydroxide has the hazard symbol . Describe the health and safety problems associated with this type of chemical. **(2 marks)**

....

....

(d) Copper oxide has the hazard symbol . Describe the health and safety problems associated with this type of chemical. **(2 marks)**

....

....

3 Describe why safety goggles should be worn as a control measure when using a substance that is corrosive. **(2 marks)**

....

....

4 Chlorine water is a toxic substance, which means it can cause harm and potentially death if swallowed, breathed in or absorbed by skin.

Describe the control measures that a person should take if they perform an experiment using a toxic substance. **(2 marks)**

....

....

Protective clothing

1 Put a cross in the box next to the correct answer below.

A scientist wears protective clothing to: **(1 mark)**

☐ **A** increase the risks identified in a scientific investigation

☐ **B** increase the hazards identified in a scientific investigation

☐ **C** reduce the risks identified in a scientific investigation

☐ **D** reduce the hazards identified in a scientific investigation.

Guided

2 The image below is of a biochemist working in a lab. He is wearing a range of protective equipment to reduce the risks associated with carrying out biochemistry experiments.

Name each item of protective clothing and explain why each is worn by this scientist. **(6 marks)**

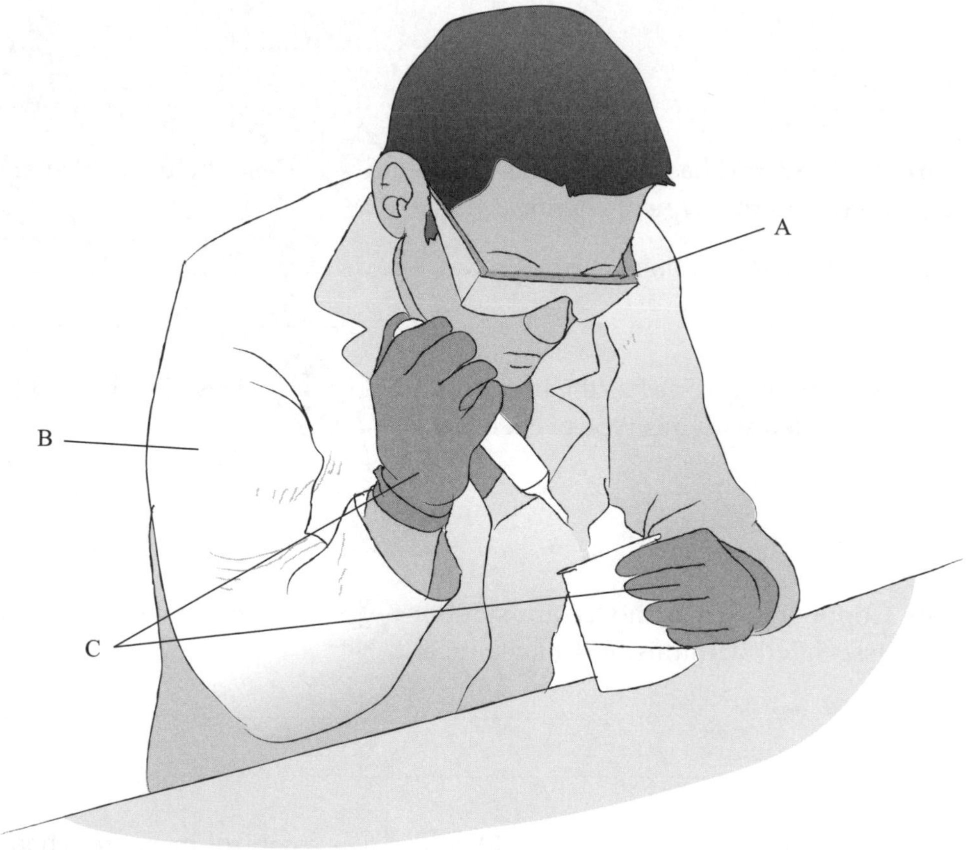

A

B

C

Item A is safety goggles. They are worn to protect the eyes from a range of hazards including splashes from corrosive or hot liquids.

....

....

....

....

Handling microorganisms

1 Put a cross in the box next to the correct answer below.

An autoclave is used to kill microorganisms using: **(1 mark)**

☐ **A** high levels of disinfectant

☐ **B** very high temperatures

☐ **C** very low temperatures

☐ **D** strong acidic solution.

2 Jill works as a microbiologist. She is investigating the effect of antimicrobial bandages on killing microbes. She places a 5 mm diameter circle of the antimicrobial bandage onto a Petri dish containing agar and a bacterial colony.

During this investigation there is the risk that Jill may transfer potentially harmful microorganisms from her hands to her mouth.

Describe **two** control measures that Jill could take to prevent this transfer from happening. **(2 marks)**

....

....

How can Jill make sure that nothing potentially harmful on her hands is transferred to her mouth?

Guided

3 A science teacher completes a risk assessment before carrying out any science experiments with their class.

Part of their risk assessment for an experiment to investigate the effect of soaps on bacterial growth is shown below.

Complete the missing pieces of information in this risk assessment. **(3 marks)**

Hazard	**Risk**	**Control measure**
Bacteria	Infection to the students carrying out the experiment.	1 Wipe down the surface with an antibacterial spray after use. 2 .... 3 ....

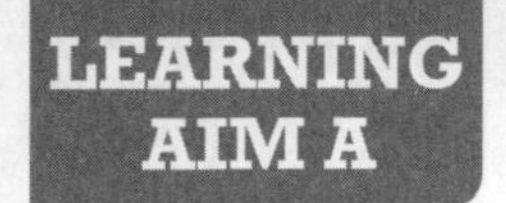

Microorganisms and wildlife

1 Complete the sentence by putting a cross in the box next to the correct answer below.

Equipment that has been used to handle microorganisms should be .... after use. **(1 mark)**

- ☐ **A** rinsed off
- ☐ **B** sterilised
- ☐ **C** thrown away
- ☐ **D** incubated.

Guided 2 Microorganisms in the air or on a person's hands could be harmful. When growing microorganisms in a Petri dish, it is important to ensure that harmful microorganisms are not transferred into the Petri dish.

Describe **two** ways of reducing the risk of growing dangerous microorganisms in a Petri dish. **(2 marks)**

The lid is taped to the dish with tape to stop ....

The Petri dish is incubated at 25 °C. ....

....

At temperatures higher than 25 °C, more rapid growth of bacteria is encouraged. This could include pathogens which are harmful to health.

3 A group of school children try to estimate how many worms are in the school playing field. They place quadrants at various points around the playing field and then pour water onto the area to encourage worms to come to the surface of the earth. The children put the worms in a bucket and then count the total number of worms that have been collected after 1 hour. The worms are then returned to the field once the experiment is over.

Guided **(a)** Describe **one** risk associated with this experiment. **(2 marks)**

The children are handling worms that have been in the soil so they may have been in contact with ....

....

(b) Describe a control measure that should be put in place to minimise this risk. **(2 marks)**

....

....

Dependent and independent variables

1 Complete the sentence by putting a cross in the box next to the correct answer.

The dependent variable is: **(1 mark)**

☐ **A** kept constant during the investigation

☐ **B** measured during the investigation

☐ **C** always changed during the investigation

☐ **D** disregarded during the investigation.

The dependent variable is the one you are investigating and recording.

2 A student is carrying out an investigation to find out how the reaction rate changes with a change in concentration of HCl. The reaction that they are investigating has the chemical equation:

$$Zn + 2HCl \rightarrow H_2 + ZnCl_2$$

(a) Identify the independent variable in this investigation. **(1 mark)**

....

(b) Identify the dependent variable in this investigation. **(1 mark)**

....

3 A sports scientist is carrying out an investigation into the effect of glucose concentrations in different sports drinks on the blood glucose concentrations of athletes who drink them. She records the blood glucose concentrations of each athlete 20 minutes after the sports drink has been consumed and then compares the results.

Guided **(a)** Describe the independent variable in this investigation. **(2 marks)**

The independent variable is the glucose concentration in the sports drink because ....

....

Guided **(b)** Describe the dependent variable in this investigation. **(3 marks)**

The dependent variable is the blood glucose concentration because ....

....

....

....

Write one more sentence on how the changes to the independent variable will affect the dependent variable

Had a go ☐ Nearly there ☐ Nailed it! ☐

Control variables

1 Complete the sentence by putting a cross in the box next to the correct answer.

A control variable is: **(1 mark)**

☐ **A** a quantity that increases

☐ **B** a quantity that remains constant

☐ **C** a quantity that decreases

☐ **D** a quantity that changes.

A control variable doesn't change.

2 A plant geneticist is trying to breed a tomato plant whose tomatoes remain firm when they are ripe. He grows two groups of 10 tomato plants: group A are genetically engineered to keep the tomatoes firm when ripe and group B is an ordinary strain of tomato plant. Both groups are grown in the same greenhouse with the same quantities of compost and water.

Guided

(a) Describe **three** control variables in this investigation. **(3 marks)**

Two of the control variables are the quantity of compost and ....

....

Another control variable is the fact that both groups are ....

....

(b) Explain why it is important to have control variables in this investigation. **(2 marks)**

....

....

....

3 Jahid is a research scientist in an oil company. He is investigating the amount of energy contained in different fractions of crude oil. He sets up a calorimeter as shown in the diagram.

He places different fractions of crude oil into the spirit lamp and burns each completely. He then measures the temperature change of the water and uses this information in the equation:

$$q = m \times C \times \Delta T$$

q = energy transferred, measured in joules (J)
m = mass of water (g)
C = specific heat capacity of water = 4.2 J/°C/g
ΔT = temperature change (°C).

Describe **two** of the control variables that should be included in this investigation and explain why they are necessary. **(4 marks)**

....

....

....

....

Measurements

Guided

1 You can convert between metric units by multiplying or dividing by 10, 100 or 1000.

Complete the table here which shows different units of length. **(8 marks)**

Size	Number of metres
1 kilometre	1000
1 metre	
100 kilometres	100 000
1 centimetre	
1 millimetre	0.001
10 millimetres	
1 nanometre	
10 centimetres	0.1

Remember 10 millimetres is the same length as 1 cm.

2 Write the following in standard form.

(a) 20 000 degrees Celsius **(1 mark)**

....

(b) One thousandth of a degree Celsius **(1 mark)**

....

(c) 150 000 000 km **(1 mark)**

....

3 The average width of a human hair is 0.000 06 m.

Write this number in standard form. **(1 mark)**

....

Units of measurement

Gravity is measured in newtons.

For each question 1–3, put a cross in the box next to the correct answer.

1 A newton is a unit of: **(1 mark)**

☐ **A** volume ☐ **B** force ☐ **C** mass ☐ **D** resistance.

2 Metres per second is a unit of: **(1 mark)**

☐ **A** power ☐ **B** area ☐ **C** velocity ☐ **D** length.

3 A millimetre is a unit of: **(1 mark)**

☐ **A** time ☐ **B** length ☐ **C** volume ☐ **D** density.

Guided

4 State the units that are used to measure density. **(1 mark)**

grams per ....

5 State the units that are used to measure power. **(1 mark)**

....

6 State the units that are used to measure resistance. **(1 mark)**

....

For each question 7–9, put a cross in the box next to the correct answer.

7 Which of the following is the symbol for cubic metres? **(1 mark)**

☐ **A** m^2 ☐ **B** m^3 ☐ **C** m^{-3} ☐ **D** m/s^3

8 Which of the following is the symbol for frequency? **(1 mark)**

☐ **A** Hz ☐ **B** Ω ☐ **C** mA ☐ **D** kJ

9 Which of the following is the symbol for current? **(1 mark)**

☐ **A** °C ☐ **B** A ☐ **C** N ☐ **D** cm

10 Mandy is a vet. She needs to be able to find the mass of the animals that she is treating in order to work out the correct dose of drugs to give them. Some of the animals that she treats are very small, such as hamsters, but the zoo animals that she looks after are very big, such as camels.

Identify the most appropriate units to measure the mass of the following animals:

(a) A hamster. **(1 mark)**

....

(b) A camel. **(1 mark)**

....

Had a go ☐ Nearly there ☐ Nailed it! ☐

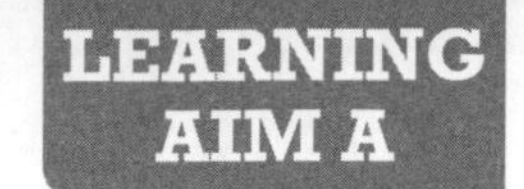

Accurate and precise measurements 1

1 The diagram shows part of a graduated cylinder. Identify the volume of liquid in this cylinder by putting a cross in the box next to the correct answer.

☐ **A** 6.7 cm^3 ☐ **B** 6.6 cm^3

☐ **C** 6.5 cm^3 ☐ **D** 6.8 cm^3 **(1 mark)**

8
7
6

When you are reading the scale on a **measuring cylinder** you should always read to the bottom of the **meniscus**.

2 Alisha is investigating how the mass of a trolley affects the speed of the trolley down a 1.5 m ramp. She uses light gates to record the times for each trial.

She takes five readings for each mass.

The results for a trolley of mass 2 kg are shown on the right.

Alisha thinks that she has gained some precise results for this part of her experiment.

Mass (kg)	Speed (m/s)
2	1.23
2	1.25
2	1.24
2	1.25
2	1.24

Guided

(a) Are these results precise? **(4 marks)**

This set of measurements is measuring the same thing and all the measurements are close together. ....

....

....

If you measure the same thing lots of times and your measurements are close together then they are **precise**.

(b) After Alisha has finished her experiment using the 2 kg mass, she realises that she had not calibrated the digital balance correctly before she weighed the trolley and the mass was actually 2.5 kg.

Explain how this error would have affected the accuracy of the results. **(2 marks)**

Accurate measurements are measurements that are close to the correct value.

....

....

(c) Explain how it is possible to have results that are precise but not accurate. **(2 marks)**

....

....

....

Had a go ☐ Nearly there ☐ Nailed it! ☐

Accurate and precise measurements 2

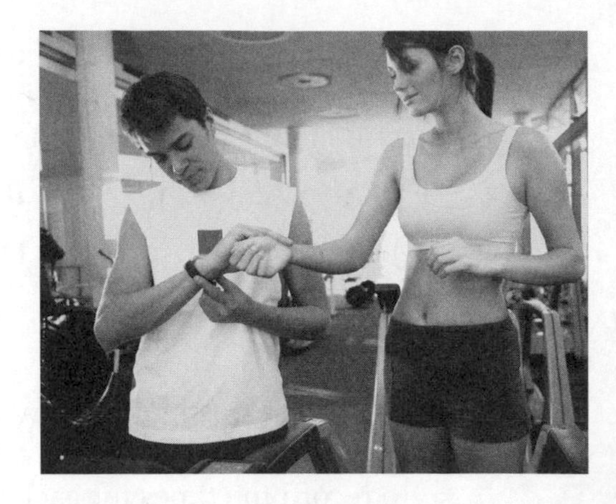

1 Jack is a personal trainer. During the initial assessment of each client he takes their resting heart rate to estimate how fit that person is. Usually, the lower the resting heart rate, the fitter the person is.

He has a new client and uses a heart rate monitor to record their resting heart rate. He also takes their pulse rate manually at the same time by taking their pulse from their radial artery at the wrist.

The results for three separate readings are shown below:

Reading number	Heart rate monitor readings (bpm)	Manual radial pulse readings (bpm)
1	65	65
2	67	66
3	66	66

(a) Calculate the mean heart rate for:

(i) the heart rate monitor readings **(1 mark)**

....

(ii) the manual radial pulse readings. **(1 mark)**

....

Always include the units of measurement in your answer – in this case it will be bpm.

Guided

(b) Explain if the manual set of radial pulse readings are accurate and precise. **(4 marks)**

The results are accurate as the manual radial pulse readings are very close or the same as

....

....

....

Range and number of measurements

1 Imran is a biomedical student. He is investigating what quantities of chemicals are most effective in making a chemical ice pack to treat sports injuries. The ice pack will contain water and ammonium chloride.

Imran uses 100 cm^3 of water for each experiment and adds different masses of ammonium chloride to the water. He takes the temperature before and after the ammonium chloride has been added and records the temperature difference for each mass. He repeats the test four times for each mass of ammonium chloride added. The results are shown in the table.

(a) Describe why the circled result is an anomalous result. **(2 marks)**

....

....

....

An **anomalous result** is a measurement that falls outside of the range of the other results or does not fit the pattern of the other results.

Mass of ammonium chloride (g)	Temperature difference (°C)			
	Test 1	Test 2	Test 3	Test 4
10	–2.1	–2.0	–2.1	–2.1
20	–3.2	–3.2	–3.1	–3.1
30	–4.5	–4.5	–4.6	–4.5
40	–5.7	–5.6	–5.7	–5.7
50	–6.2	(–6.9)	–6.2	–6.2
60	–7.1	–7.1	–7.0	–7.0

(b) Explain whether a sufficient number of tests have been carried out for each mass of ammonium chloride. **(3 marks)**

....

....

....

(c) Explain whether enough tests have been carried out with the independent variable. **(3 marks)**

....

....

....

Guided

(d) Name and describe the type of reaction that takes place which makes the chemical pack go cold. **(2 marks)**

The reaction is an .... reaction, which means that heat energy is .... This means that the contents of the packet ....

Use scientific terminology that you learned in Unit 5 Applications of chemical substances, to help you to answer this question.

Had a go ☐ Nearly there ☐ Nailed it! ☐

Writing a method

Guided

1 Describe the details that a scientific method should include.

The method for an investigation should include: **(4 marks)**

- a list of ....
- how much ....
- ....

2 A gardener wants to find out if fertiliser with added nitrogen helps to increase sunflower growth. She plans to compare plants grown without nitrogen fertiliser to plants grown with nitrogen fertiliser. She will measure the increase in height of each plant during the experiment.

The method for this experiment is shown below.

Equipment: eight small plant pots with holes in the bottom, potting soil, one packet of sunflower seeds, non-nitrogen fertiliser, nitrogen fertiliser mix for house plants, empty glass or plastic bottle for mixing nitrogen fertiliser, digital scales, beaker, labels.

Method

1 Using the digital scales, measure 500 g of potting soil into each of the pots.
2 Label four pots with 'Nitrogen F' and four pots with 'No Nitrogen'.
3 Add 50 g of nitrogen fertiliser to the 'Nitrogen F' pots and 50 g of non-nitrogen fertiliser to the 'No Nitrogen' pots.
4 Moisten the soil in each pot with 100 cm^3 of water.
5 Plant two seeds in each pot according to the seed packet instructions.
6 Place all eight pots near a sunny window.
7 If both seeds germinate, remove the smaller seedling to leave one plant.
8 Check the plants regularly and water every second day. All plants should receive the same amount of liquid at each watering.
9 Every week measure the height of each plant using a ruler and record these results.

(a) Identify **one** piece of equipment that is missing from the equipment list. **(1 mark)**

....

(b) Does the method provide details of the quantity and types of substances required? Explain your answer. **(2 marks)**

....

....

(c) Describe **two** control variables used in this plan. **(2 marks)**

....

....

A **control variable** is a variable that remains constant.

Hypotheses

1 An investigation was carried out using two rose plants. Both plants received yellow light. One plant was kept at 20 °C and the other was kept at 15 °C. Other than the different temperatures, all other conditions were identical.

(a) State **one** possible hypothesis that could be tested in this investigation. **(1 mark)**

....

A **hypothesis** states clearly what you expect to happen in the investigation and is based on relevant scientific ideas.

(b) What data would need to be collected in order to test the hypothesis stated in part **(a)**? **(1 mark)**

....

2 A student squeezed a hand grip dynamometer to test for hand grip strength. They repeated this test every minute for a 10 minute period. The data obtained are shown in the table below.

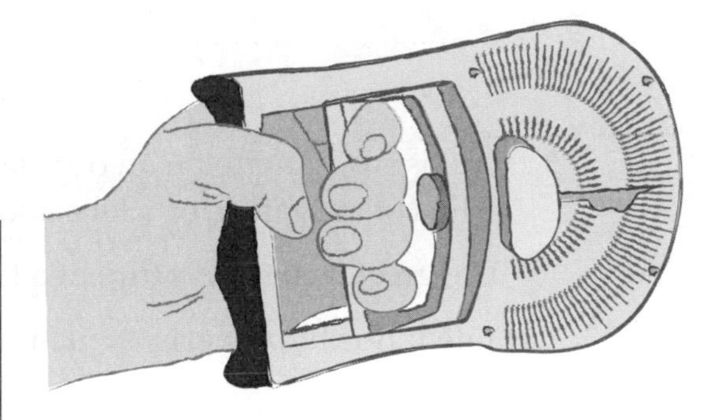

Trial	Hand grip dynamometer reading (kg)
1	32
2	29
3	28
4	27
5	26
6	25
7	23
8	21
9	19
10	17

State **one** hypothesis that these data would support. **(1 mark)**

....

Guided

3 Chloe has read that if a person has a cold, the more vitamin C a person consumes the less time they will have the cold symptoms. She wants to carry out an experiment to test this.

State an appropriate hypothesis for this experiment. **(1 mark)**

The more vitamin C a person with a cold consumes ....

....

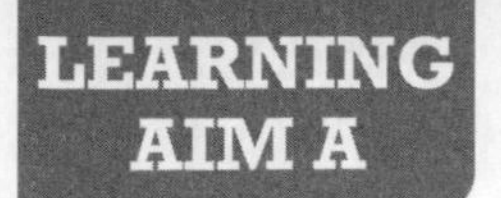

Had a go ☐ Nearly there ☐ Nailed it! ☐

Qualitative and quantitative hypotheses

1 Sharon writes the following hypothesis:

'Suncreams with an SPF factor of 30 will be twice as effective at blocking UV radiation from sunlight as suncreams with an SPF factor of 15.'

Is this hypothesis quantitative or qualitative? **(1 mark)**

....

2 Explain the difference between a quantitative and a qualitative hypothesis. **(2 marks)**

....

....

....

3 Raul is studying to be an optician. He wants to investigate the relationship between the radius of curvature of a lens and its focal length. He writes the hypothesis:

'The greater the curvature of a lens, the larger the focal length.'

State if this hypothesis is qualitative or quantitative. **(1 mark)**

....

Guided 4 A dietician is providing daily eating plans for female adults who are at least 30 kg overweight. They plan to carry out an investigation to find out whether, if these females consume 1500 kcalories a day, they will lose at least 1 kg of body mass per week.

Write a quantitative hypothesis for this investigation. **(2 marks)**

If females who are 30 kg or more overweight consume ....

....

....

Planning an investigation

Guided

1 A mathematician has made a claim that people who take part in vigorous exercise for 15 minutes or more every day are able to solve maths problems more rapidly compared to people who do not take part in daily vigorous exercise.

Write a plan for a controlled experiment that could be conducted to test this claim. **(6 marks)**

In your answer, make sure you:
- state the purpose of the experiment
- make a prediction about what you think will happen
- describe how the experimental group will be treated and how the control group will be treated
- state what data will be collected during the experiment.

Aim: The purpose of this experiment is ....

....

Hypothesis: ....

....

Method: There will be two groups in the experiment – groups A and B.

Group A is the ....

....

Group B is the control group and will not ....

Both groups will be given the same mathematical problems and ....

....

....

If group A solves the maths problems more rapidly than the control group then ....

....

Learning aim A sample questions

A pharmaceutical technician in a drugs company is carrying out an experiment to find out how effective a new brand of antibiotic is at preventing bacterial growth.

Bacteria are spread on the surface of 5 different agar plates. She places a small paper disc containing a different concentration of the new antibiotic on each of these plates. They are then placed in an incubator at 25 °C.

After 48 hours she removes the agar plates from the incubator. She measures the diameter of the exclusion zone on each plate where there are no bacteria.

After another 48 hours she repeats the experiment with the same concentrations of antibiotic as the first experiment, to see if the exclusion zones on the agar plates are the same.

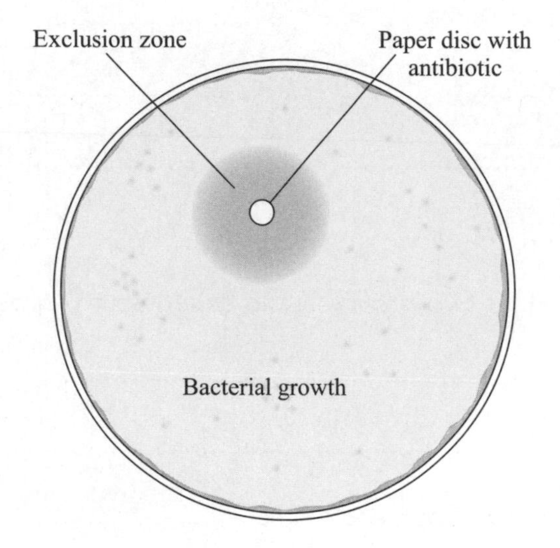

1 **(a)** Identify the dependent and independent variable in this investigation. **(2 marks)**

....

....

(b) Describe **two** variables that need to be controlled to make this a fair test. **(4 marks)**

....

....

....

....

Tables of data

1 Complete the sentence by putting a cross in the box next to the correct answer.

When placing data in a table the first column should contain the: **(1 mark)**

☐ **A** control variable ☐ **B** dependent variable

☐ **C** independent variable ☐ **D** units.

2 Kevin works as a design engineer on the tyres of Formula 1 cars. He wants to test how different materials in tyres can increase the length of time that they can be used in races.

He uses one car and drives it around a test track 10 times, then takes the tyre tread depth. He repeats this process with five different types of tyres, which are given the names, A, B, C, D and E. All of the tyres start with a tread depth of 1.70 mm.

The data from his test are shown below:

Tyre A – 1.57 mm, Tyre B – 1.69 mm, Tyre C – 1.35 mm, Tyre D – 1.23 mm, Tyre E – 1.59 mm.

Put these data into the correct columns in the table. **(4 marks)**

Tyre	Tread depth (mm)

3 Mary is carrying out an experiment to find out how the speed of a person running 800 m changes over time.

(a) Which of these variables is the independent variable? **(1 mark)**

....

(b) Which column should the independent variable be placed in a table? **(1 mark)**

....

Guided

4 Bill carries out an experiment to find out how the temperature of a hand warmer changes over a 6 minute period of time.

His results are shown in the table on the left. The data have not been displayed correctly. Put the hand warmer data into the table on the right to clearly show the trend in the data. **(3 marks)**

Temperature (°C)	Time (min)
16	0
24	2
18	1
28	3
33	5
33	6
33	4

Time (min)	Temperature (°C)
0	16
6	33

Had a go ☐ Nearly there ☐ Nailed it! ☐

Identifying anomalous results from tables

1 Complete the sentence by putting a cross in the box next to the correct answer.

An anomalous result can also be called: **(1 mark)**

☐ **A** a variable ☐ **B** a control ☐ **C** a precaution ☐ **D** an anomaly.

2 Kian is a school student carrying out a science experiment. He cuts out 20 pieces of potato which are all the same diameter and length. He places five pieces into each of four beakers containing different concentrations of sugar solutions. He measures each potato piece again after 24 hours.

The results are shown in the table below.

Sugar solution concentration (g/l)	Original length of potato (mm)	Length after 24 hours (mm)
0	50	52, 51, 51, 50, 51
5	50	44, 43, 42, 44, 43
8	50	43, 42, 44, 51, 43
10	50	42, 41, 40, 41, 40

(a) Calculate the mean length of the potato after 24 hours for each sugar solution concentration. **(4 marks)**

Sugar solution concentration (g/l)	Mean length after 24 hours (mm)
0	
5	
8	
10	

If you cannot explain why there is an anomalous result then it should be ignored and left out when calculating the mean of the results.

(b) Identify which of these results is an anomalous result. **(1 mark)**

....

Guided

(c) Explain your answer to part **(b)**. **(3 marks)**

This is an anomalous result because it falls outside the range of the other results in the group at 8 g/l sugar solution. ....

....

....

Write two more sentences comparing the anomalous result to the range of the other results, and explaining how it doesn't fit the pattern.

Identifying anomalous results from graphs

1 Marios carries out an investigation into the effect of adding different masses of ammonium chloride to 30 cm^3 of water.

The results of the investigation are shown below:

Mass 3 g, temperature rise 4 °C; mass 6 g, temperature rise 6 °C; mass 9 g, temperature rise 7 °C; mass 12 g, temperature rise 12 °C; mass 15 g, temperature rise 11 °C.

(a) Complete the table below to display these results. **(3 marks)**

Mass of ammonium chloride (g)	3				
Temperature rise (°C)	4				

(b) Plot a graph of the data on the grid below. **(6 marks)**

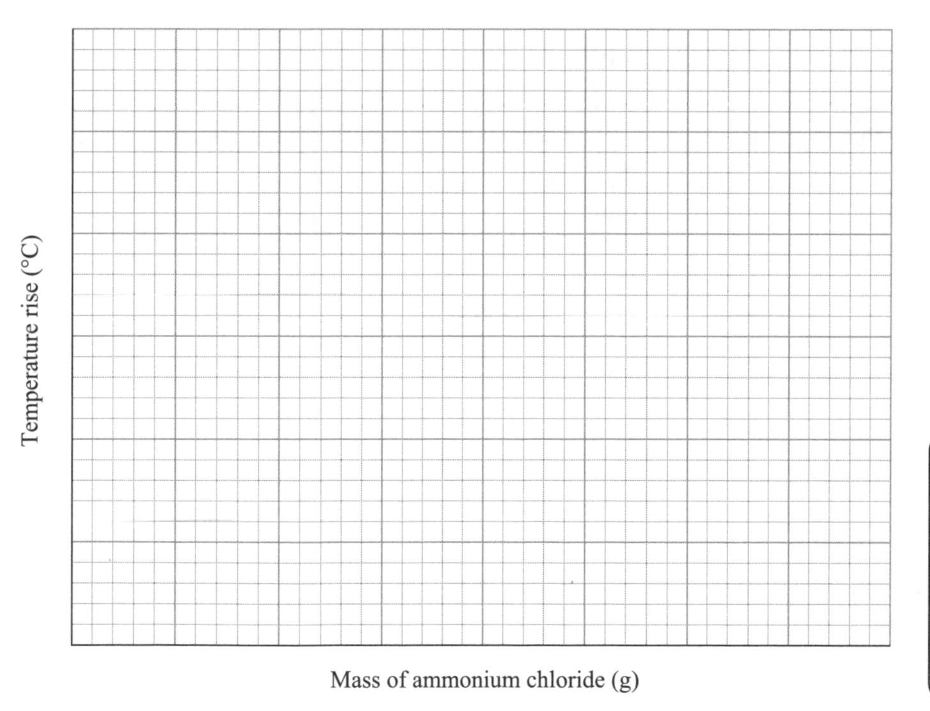

An anomalous result on a graph will fall above or below the line of best fit as it does not follow the pattern of the rest of the results.

(c) Circle the anomalous data on the graph. **(1 mark)**

Guided

(d) Explain your answer to part **(c)**. **(3 marks)**

This is an anomalous result because ....

....

The pattern is a steady increase in ....

....

At 12 g of ammonium chloride added ....

....

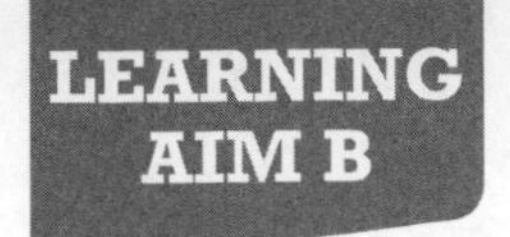

Had a go ☐ Nearly there ☐ Nailed it! ☐

Excluding anomalous results

1 Which of the following steps should be taken if an anomalous results is found in an experiment? Complete the sentence by putting a cross in the box next to the correct answer. **(1 mark)**

The measurement should be:

☐ **A** assumed to be accurate

☐ **B** included in calculations

☐ **C** reliable

☐ **D** ignored.

2 Simon works in a brewery. One of his roles is to monitor the ethanol percentage in the beers that the brewery makes. He takes 15 different samples from different parts of the fermentation tank and measures the percentage of ethanol in each. He then calculates the mean of the results to determine the actual percentage of ethanol for that batch of beer.

The results of his tests are shown in the table below.

Sample number	Ethanol (%)
1	3.51
2	3.49
3	3.49
4	3.51
5	3.49
6	3.50
7	3.51
8	3.50

Sample number	Ethanol (%)
9	3.51
10	3.49
11	3.68
12	3.50
13	3.49
14	3.50
15	3.49

(a) Which of these percentages is an anomalous result? **(1 mark)**

....

(b) Describe how this anomalous result should be dealt with when working out the mean percentage. **(1 mark)**

....

(c) Calculate the mean ethanol percentage in this batch of beer. **(2 marks)**

....

The mean is the total of all the values divided by the number of values.

Calculations from tabulated data

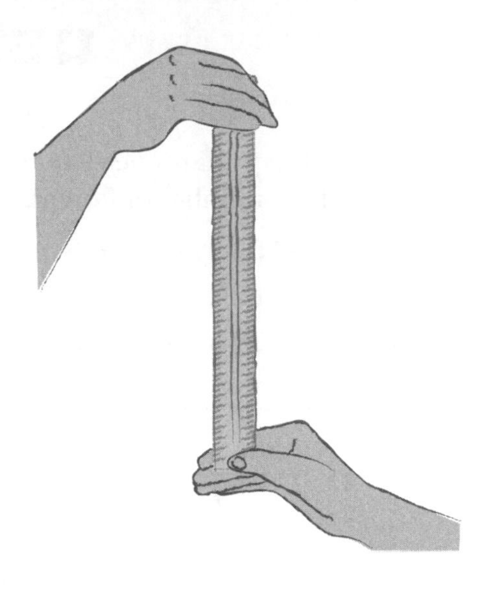

1 A sports psychologist is carrying out an investigation into the effects of temperature on a person's reaction times. She carries out a ruler drop experiment that tests reaction times with five males aged between 18 and 19 years.

For the first test, the person places their hands in a water bath at a temperature of 40 °C for 2 minutes. They then carry out the reaction test three times and each score is recorded. For the second test, the person cools their hands in a water bath at 5 °C for 2 minutes, then repeats the reaction test a further three times.

The results are shown in the table below.

Guided

(a) Calculate the means to the nearest 100th of a second. **(10 marks)**

Person	Test results after heating hands in a water bath at 40 °C (seconds)	Mean test result after heating (seconds)	Test results after cooling hands in a water bath at 5 °C (seconds)	Mean test result after cooling (seconds)
1	0.42, 0.40, 0.41	0.41	0.48, 0.46, 0.46	
2	0.52, 0.56, 0.53		0.52, 0.56, 0.59	
3	0.39, 0.36, 0.33		0.42, 0.42, 0.43	
4	0.36, 0.34, 0.33		0.41, 0.41, 0.42	
5	0.55, 0.50, 0.49		0.60, 0.59, 0.59	0.59

(b) Work out the mean reaction rates to the nearest 100th of a second for the whole group after:

(i) heating the hands in a water bath **(1 mark)**

....

(ii) cooling the hands in a water bath. **(1 mark)**

....

Calculations from tabulated data – using equations

1 A landscape gardener wanted to find out whether adding compost to rose plants would affect the number of flowers that the roses produced.

She used twelve separate pots, 6 with compost and 6 without, and planted one rose seedling into each pot.

After 8 weeks she counted how many buds were on each rose plant.

Number of rose flowers without compost	Number of rose flowers with compost
6	7
4	9
5	9
4	11
8	8
3	6

Calculate the mean number of rose flowers for plants grown:

Guided

(a) without compost **(1 mark)**

6 + 4 + 5 + 4 + 8 + 3 = ...

Number of rose plants = ...

Total number of flowers divided by number of plants = ... ÷ .... = ...

(b) with compost. **(1 mark)**

....

2 The table below shows results from a trolley experiment using light gates to record the time taken for the trolley to travel between points A and B.

Total distance travelled between A and B	5.8 cm
Total time taken between A and B	0.6 s

Calculate the mean speed of the trolley. **(2 marks)**

Speed = distance / time
The units of speed are m/s.

....

Significant figures

1 Complete the sentence by putting a cross in the box next to the correct answer.

2.567 cm rounded to two significant figures is: **(1 mark)**

☐ **A** 2.6

☐ **B** 3.0

☐ **C** 2.5

☐ **D** 2.4

Significant figures are the digits included in a measurement.

2 A town has a population of 15 645 people.

State this number to:

(a) two significant figures **(1 mark)**

....

The first two figures are 15. However, the number after 5 is higher than 5 so the number needs to be rounded up.

(b) three significant figures **(1 mark)**

....

(c) four significant figures. **(1 mark)**

....

Guided

3 Penny is a nurse. She uses digital bathroom scales to measure the mass of a person in order to calculate their BMI.

She records different values of mass in kg for five different people – the results are shown in the table below.

Complete the table by writing these masses to two significant figures. **(5 marks)**

Person	Mass (kg)	Mass to two significant figures (kg)
1	62.34	62
2	79.81	
3	75.50	
4	68.29	
5	55.61	

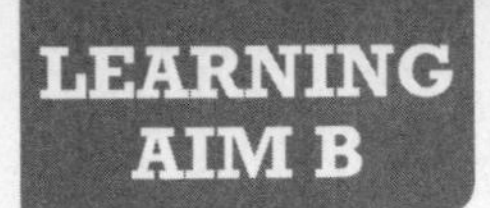

Bar charts 1

1 Complete the sentence by putting a cross in the box next to the correct answer.

When drawing a bar chart, the dependent variable should be: **(1 mark)**

☐ **A** plotted on the y-axis ☐ **B** plotted on the x-axis

☐ **C** excluded from the chart ☐ **D** a qualitative value.

The dependent variable is the outcome that is measured during the investigation.

Guided

2 Identify **six** points to remember when drawing a bar chart. **(6 marks)**

The chart should have a title that provides a summary of the chart content.

The names of the variables should be written on each axis.

....

....

....

....

3 Zenyap is investigating the pH of substances that can be found in the home. The results of this investigation are shown in the table.

(a) Plot these results as a bar chart on the grid below. **(6 marks)**

Substance	pH
lemon juice	2
orange juice	3
water	7
indigestion liquid	9
oven cleaner	14

(b) Do you think a bar chart is a good way to present these data? Explain your answer. **(2 marks)**

....

....

Bar charts 2

Guided

1 A nurse investigates the rate at which analgesics relieve headaches. The results of his study are shown in the bar chart below.

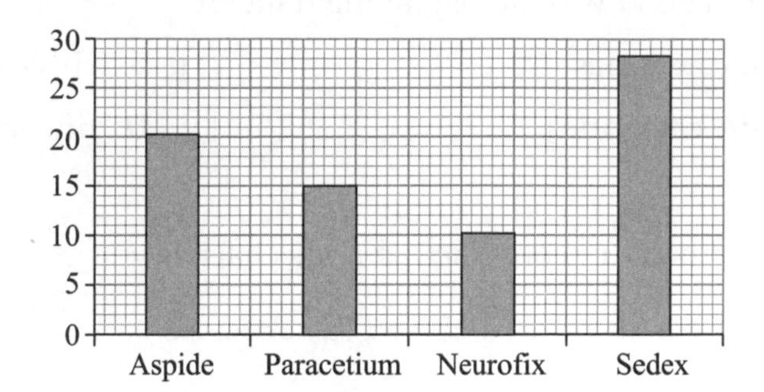

There should be three features and three explanations in your answer.

What features are missing from this bar chart? Explain your answer. **(6 marks)**

There is no title. This should be included to provide a summary of the chart content.

....

....

....

2 A cardiac surgeon was interested in finding out how Body Mass Index affects the chances of a person's survival after having a heart attack.

The results of a study of 100 people are shown in the table below.

Heart attack survivors alive after 5 years (%)	Body Mass Index rating
23	Underweight
55	Recommended weight
16	Overweight
6	Obese

Plot these data on a bar graph.
(6 marks)

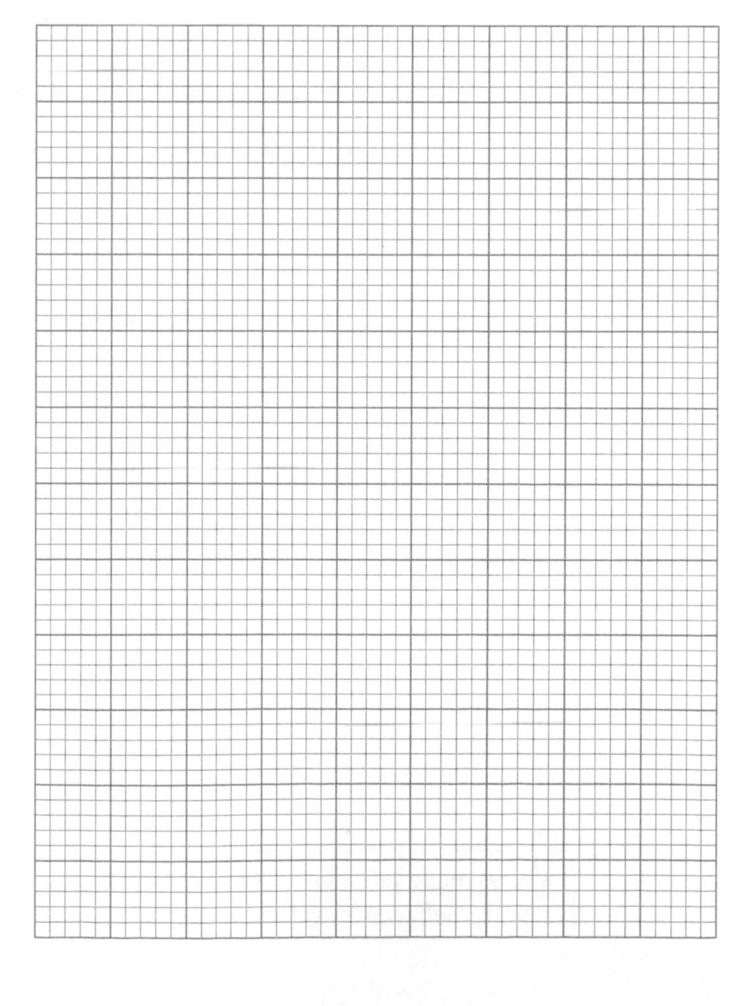

Had a go ☐ Nearly there ☐ Nailed it! ☐

Pie charts

1 Complete the sentence by putting a cross in the box next to the correct answer.

A pie chart is a good way of displaying data for: **(1 mark)**

☐ **A** eight categories ☐ **B** six categories

☐ **C** twelve categories ☐ **D** fifteen categories.

Guided

2 Describe the process of constructing a pie chart. **(4 marks)**

To construct a pie chart you need to calculate the ....

which have to add up to 360°.

Work out the total number in the sample.

Divide the number in each category by ....

....

Use a compass to ....

....

3 A forensic scientist wants to display the fingerprint data from a crime scene on a pie chart.

The results of the different fingerprint types from the crime scene are shown in the table opposite.

Draw a pie chart to display this information.

Show all of your calculations. **(6 marks)**

Fingerprint type	Prints (%)
Loops	65
Arches	10
Whorls	25

4 The pie chart opposite has been drawn to show the shoe sizes of a class of twenty-five Year 5 school children.

Describe how this pie chart could be improved. **(2 marks)**

....

....

....

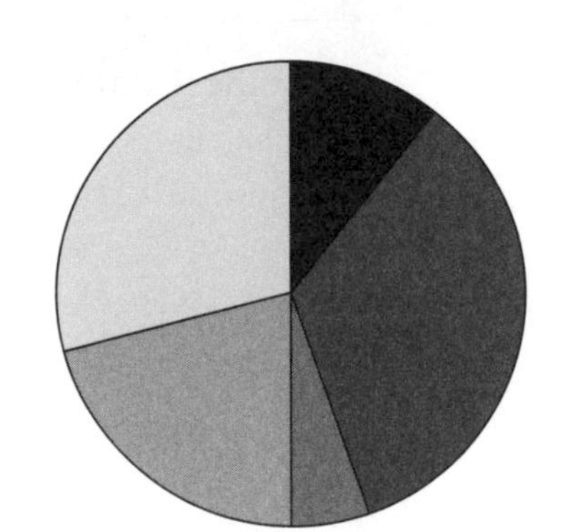

Always make sure a pie chart has a key so it is clear what each sector represents.

Line graphs 1

1 Reena is investigating the solubility of sodium chloride in water at different temperatures.

She uses the same quantity of water at each temperature and adds sodium chloride to each, stirring the solution until no more will dissolve. The results of her investigation are shown in the table.

Plot a line graph of this data using the graph paper below. **(6 marks)**

Water temperature (°C)	Quantity of solute added to 500 g water (g)
0	18
20	20
40	24
60	29
80	36
100	49

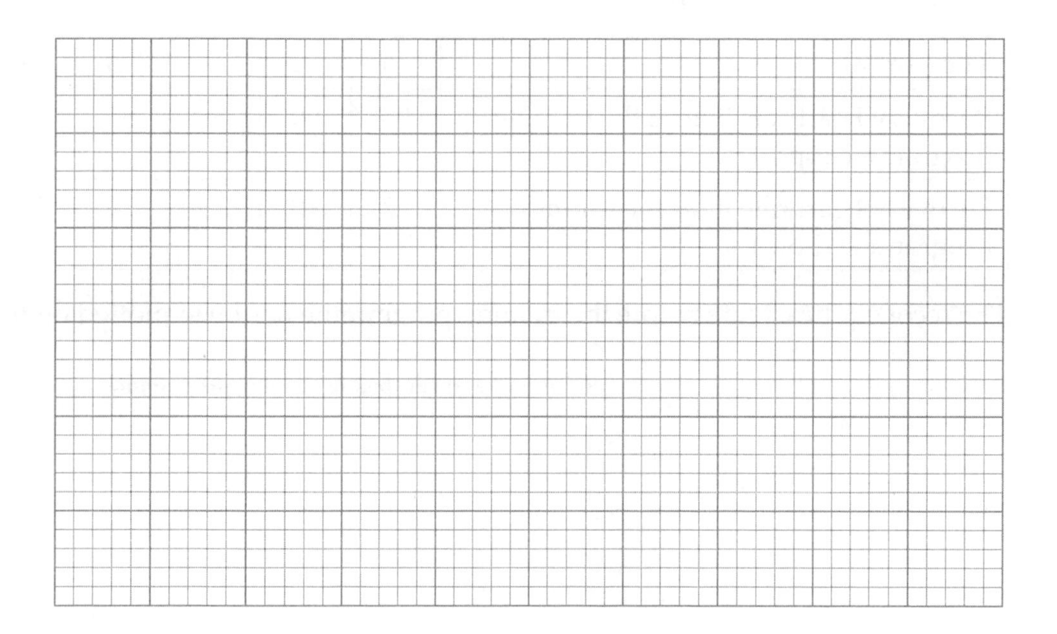

- Make sure you mark an appropriate scale on the axes.
- The temperature is the independent variable so that should be plotted on the *x*-axis.
- Always give the graph a title so that it is clear what data is being displayed.

2 A marathon runner had their core temperature monitored during the 26.2 mile race. The results are shown on the graph.

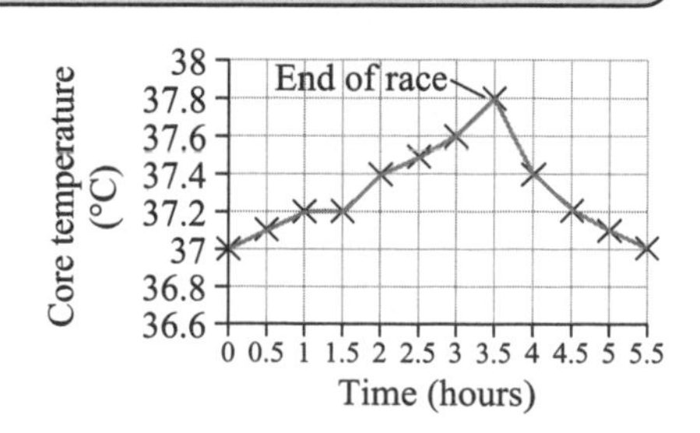

(a) What was the increase in core temperature over the course of the race? **(1 mark)**

....

(b) How long did it take for the marathon runner's core temperature to return to normal after the race? **(1 mark)**

....

Had a go ☐ Nearly there ☐ Nailed it! ☐

Line graphs 2

1 Complete the sentence by putting a cross in the box next to the correct answer.

Line graphs are normally used when the independent variable is quantitative and the dependent variable is: **(1 mark)**

☐ **A** qualitative ☐ **B** quantitative ☐ **C** controlled ☐ **D** constant.

2 John is trying to improve his cardiovascular health. He knows that after taking part in exercise the faster his heart rate returns to resting levels, the fitter he has become.

He takes part in a 12-week training programme and compares his post-exercise heart rate before and after the training programme.

The results are shown below.

Time (min)	**0**	**2**	**4**	**6**	**8**	**10**	**12**
Recovery heart rate before the training programme (bpm)	95	90	86	80	75	72	72
Recovery heart rate after the training programme (bpm)	102	90	78	75	72	70	70

The recovery heart rate before the training programme has been plotted on the graph below.

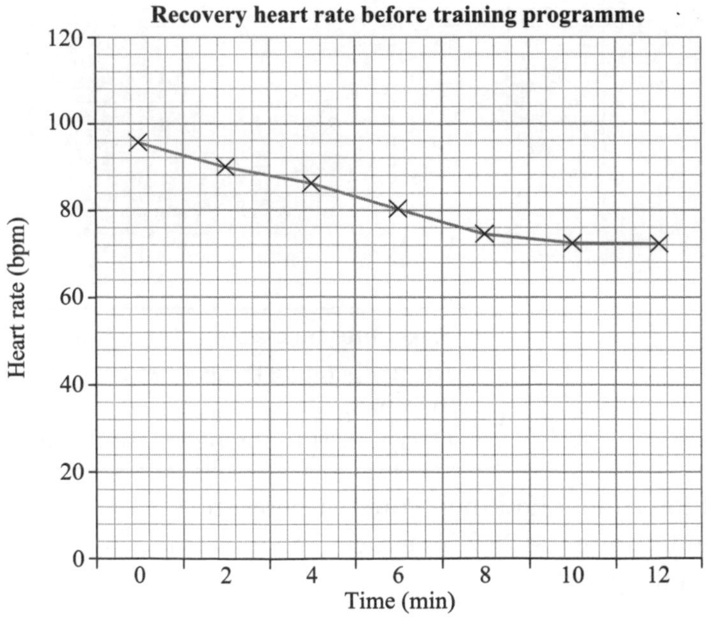

(a) Plot the recovery heart rate after the training programme on this graph. **(1 mark)**

Guided

(b) Compare the trends in these different sets of data. **(3 marks)**

Both heart rates decrease after having stopped exercising. ....

....

....

Write another sentence to compare the decrease in heart rates before and after the training programme.

Straight lines of best fit

1 Complete the sentence by putting a cross in the box next to the correct answer.

A line of best fit is used to show: **(1 mark)**

☐ **A** anomalous data ☐ **B** reliable data

☐ **C** changes in data ☐ **D** a trend in data.

2 Marty carried out an experiment to find out how the length of a spring was affected by adding weights to it. Different weights were attached to a suspended spring and the resulting total length of the spring was measured.

The data for this experiment are shown in the table below.

Attached weight (N)	Total spring length (cm)
0.98	0.37
1.96	0.42
2.94	0.51
3.92	0.59
4.91	0.64

Guided

(a) Plot a graph of these data on the grid below. **(6 marks)**

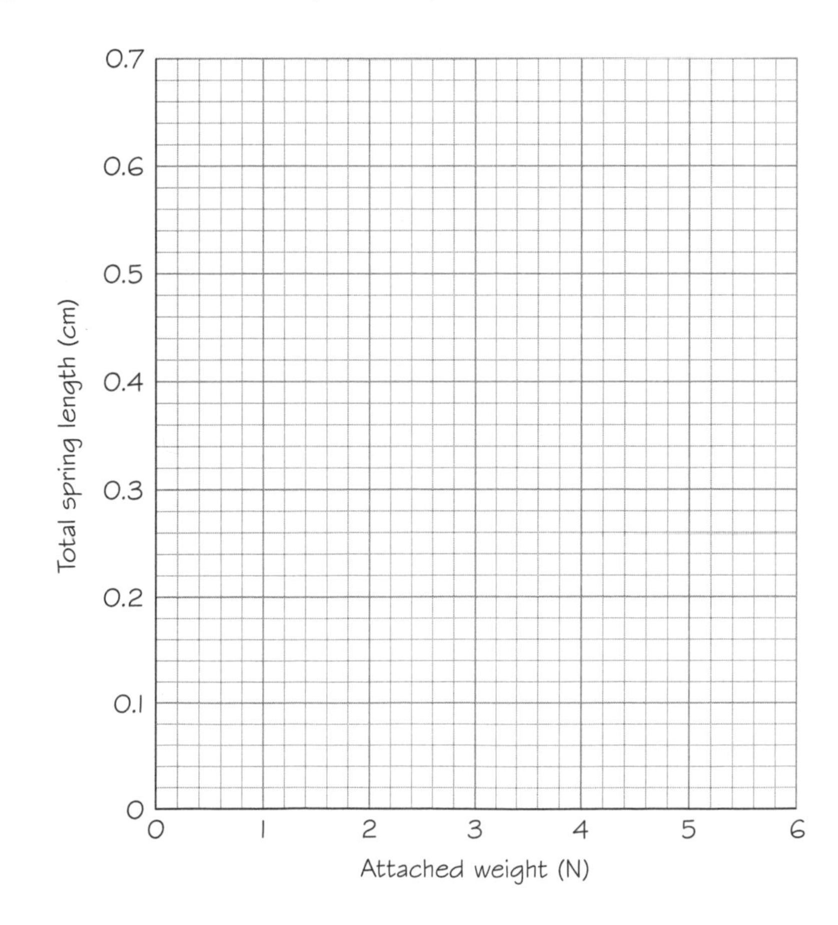

(b) Draw the line of best fit. **(2 marks)**

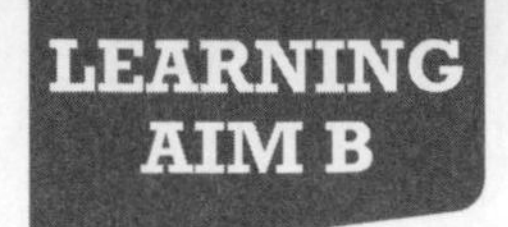

Curves of best fit

1 Complete the sentence by putting a cross in the box next to the correct answer.

When drawing a line of best fit, you should: **(1 mark)**

☐ **A** draw the line so it passes through every data point

☐ **B** exclude any anomalous data

☐ **C** draw the line so it does not pass through any data points

☐ **D** not use a ruler.

Guided

2 How would you best show a pattern of results when only two or three points on a graph lie close to a straight line? **(2 marks)**

When very few points on a graph seem to lie close to a straight line, the best way to show the pattern of results is to ....

3 Lottie is investigating how cycling down a hill affects the distance covered by a cyclist.

She attaches a device to her bicycle that measures the time and distance travelled. She then cycles for 800 m. The first 100 m is on a flat road and then the rest of the distance is down a hill with a steady decline.

The results of the investigation are plotted on the graph below.

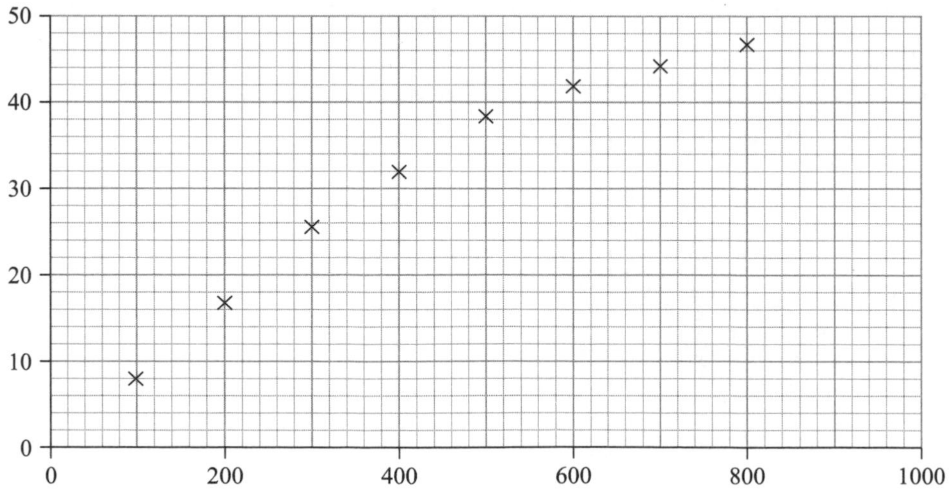

(a) Complete the graph by adding a title and labelling the axes. **(3 marks)**

(b) Draw the curve of best fit for these data. **(2 marks)**

Selecting types of graph 1

1 Put a cross in the box next to the correct answer.

It is necessary to use a protractor to draw this type of graph: **(1 mark)**

☐ **A** bar chart ☐ **B** line graph ☐ **C** pie chart ☐ **D** curve of best fit.

2 Sharon uses a digital pH meter to record the pH of different substances that can be found in the home.

The results of her investigation are shown in the table.

What type of graph would be most appropriate to display these data? Explain your answer. **(3 marks)**

Substance	pH
skimmed milk	6.4
bottled water	7.1
lemon juice	2.3
tomato juice	4.3
vinegar	3.3
toilet cleaner	11.5

....

....

....

Guided

3 Imran adds 100 cm^3 of water to ammonium chloride. He takes the temperature before and after the ammonium chloride has been added and records the temperature difference for each mass.

The results of the experiment are shown here.

Mass of ammonium chloride (g)	10	20	30	40	50	60
Temperature difference (°C)	−2.1	−3.2	−4.5	−5.7	−6.2	−7.1

(a) Plot these results on the graph paper. Draw a line of best fit through the points. **(6 marks)**

(b) Explain why this type of graph is the best way to present these data. **(3 marks)**

Both sets of variables are .... data so a .... graph is best. As the dependent variable is increasing in line with ....

....

....

....

....

Had a go ☐ Nearly there ☐ Nailed it! ☐

Selecting types of graph 2

1 The Earth's atmosphere is made up of three main gases: nitrogen, oxygen and argon. The percentage of these gases is shown in the table on the right.

Present these data in the most appropriate way, using a graph or chart. **(6 marks)**

Gas	In dry air (%)
nitrogen	78
oxygen	21
argon	1

2 A biologist carried out an investigation into the pH of water in ponds at different distances away from a city centre. The mean results of the investigation are shown in the table on the right.

(a) Plot these data on an appropriate graph. **(6 marks)**

Distance away from city centre (km)	Mean pH of pond water
5	6.5
10	6.7
15	6.7
20	6.9
25	6.9
30	7

The sets of data are all quantitative so the most appropriate type of graph will be a line graph.

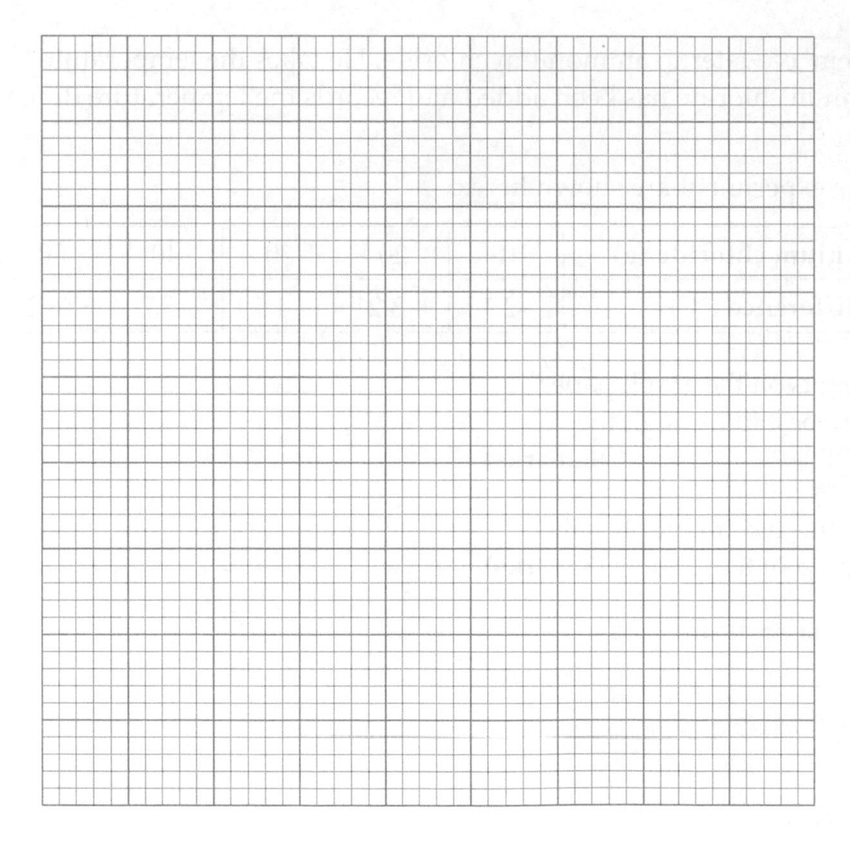

Guided

(b) Describe the trend shown by this data. **(2 marks)**

The further away from the city centre, the ....

....

(c) Should a line of best fit should be added to this graph? Explain your answer. **(2 marks)**

....

....

Finding values from graphs 1

1 Julia is investigating the temperature rise that occurs when magnesium powder is added to copper sulfate solution.

A table of these results is shown here.

Mass of magnesium powder (g)	Temperature rise (°C)
0.0	0
0.2	12
0.4	20
0.6	28
0.8	37

Guided

(a) Plot these results on the grid below. **(6 marks)**

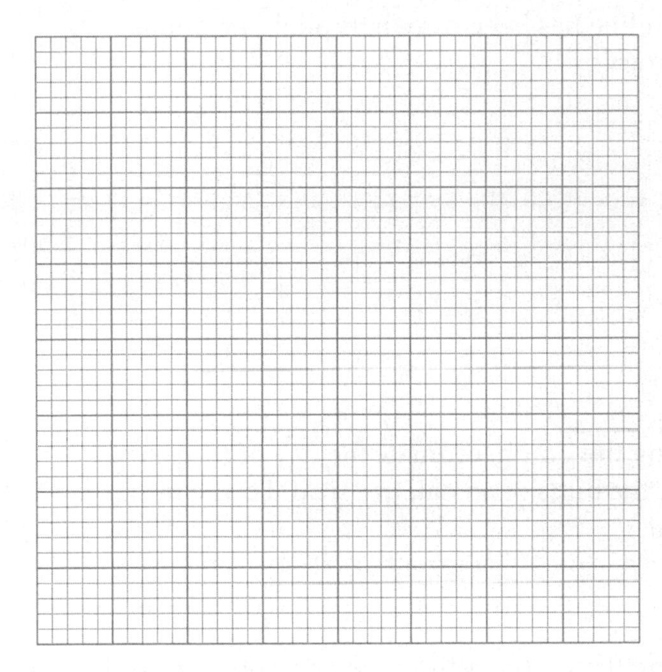

(b) Draw a best fit line on the graph. **(1 mark)**

(c) Use the graph to estimate the temperature rise if:

(i) 0.3 g of magnesium was added to the copper sulfate solution **(1 mark)**

....

(ii) 0.7 g of magnesium was added to the copper sulfate solution. **(1 mark)**

....

Draw a line vertically up from the *x*-axis at the value of the independent variable that you are interested in, then when this meets the best-fit line draw a line horizontally across to the *y*-axis.
Read the value of the dependent variable from the scale.

Had a go ☐ Nearly there ☐ Nailed it! ☐

Finding values from graphs 2

1 Complete the sentence by putting a cross in the box next to the correct answer.

Using a line of best fit to estimate values that are outside the range of values used in an investigation is called: **(1 mark)**

☐ **A** extrapolation ☐ **B** extended ☐ **C** extenuation ☐ **D** exacerbation.

2 A UK-based environmental scientist is investigating the effects of climate change on oak trees. He has used data that were taken every year from 1950 which recorded the date of buds bursting on oak trees. He plotted this information on a graph. He found that over the last 60 years the oak tree leaves are starting to come out earlier each year.

A best fit line has been drawn through the points on the graph.

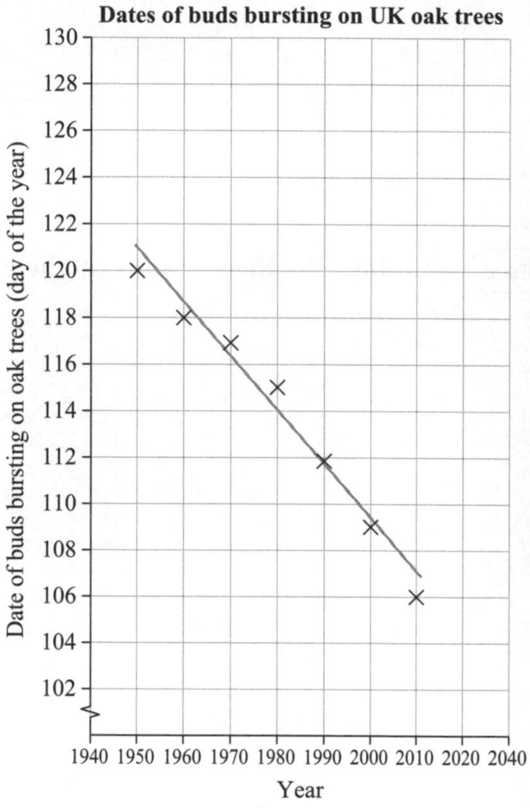

(a) Using the graph, estimate the day of the year that the buds of an oak tree would burst in 2020. **(1 mark)**

....

(b) Using the graph, estimate the day of the year that the buds of an oak tree would have burst in 1940. **(1 mark)**

....

(c) Using the graph, estimate the day of the year that the buds of an oak tree would have burst in 1995. **(1 mark)**

....

Guided

(d) Describe why it is only possible to estimate these values from the graph. **(2 marks)**

You can only estimate these values as the extrapolation of the line is based on ....

....

This trend may not continue in line with ....

Calculating gradients from graphs

1 Complete the sentence by putting a cross in the box next to the correct answer.

The gradient on a graph of distance against time can be used to work out: **(1 mark)**

☐ **A** power ☐ **B** force ☐ **C** acceleration ☐ **D** speed.

Guided

2 Anya carries out an investigation into the relationship between voltage and current in a circuit. The graph of her results can be used to work out the resistance for a resistor at a fixed temperature in the circuit. Complete the calculations to find the answer. **(3 marks)**

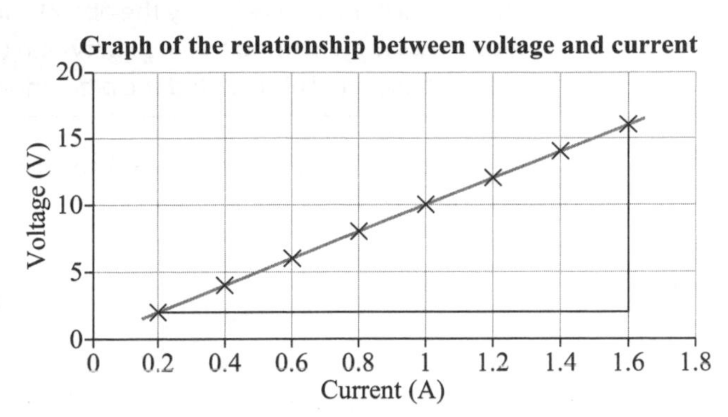

Change in y-axis values = 16 − 2 = 14

Change in x-axis values = 1.6 − 0.2 = ...

Gradient = change in y-axis value/change in x-axis value

= 14/................ = ...

For a resistor at fixed temperature, this graph is a straight line and the gradient is the resistance in ohms (Ω).

3 Hussain is investigating the speed of a two-wheeled electronic scooter.

The results of this investigation are shown below.

Distance (m)	Time (s)
0	0
20	2.55
40	4.56
60	8.00
80	9.91
100	11.23
120	14.21
140	15.81

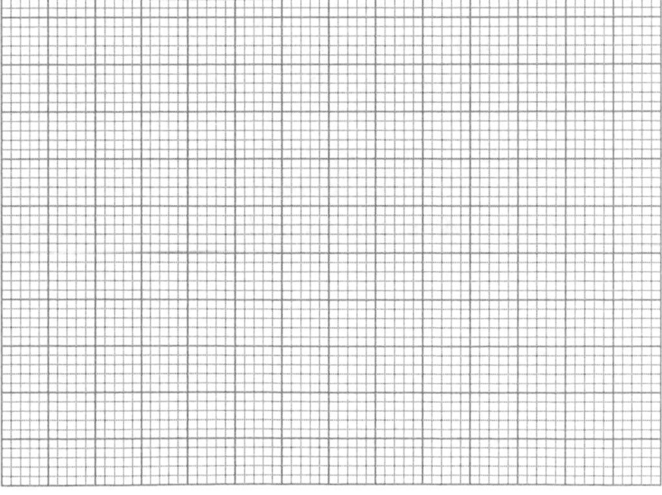

(a) Plot these results on the grid. **(6 marks)**

(b) Draw a line of best fit. **(1 mark)**

(c) Calculate the gradient of the line to work out the speed of the scooter over the 140 m track. **(2 marks)**

....

Had a go ☐ Nearly there ☐ Nailed it! ☐

Calculation of areas from graphs

1 Sian investigates the speed of a trolley over a period of time. The results of her investigation are shown on the graph opposite.

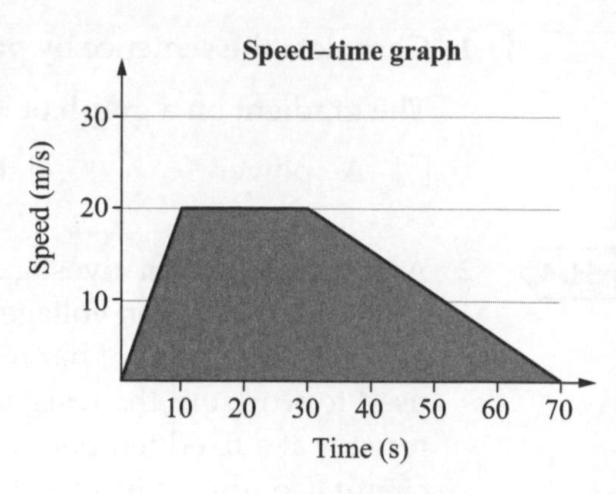

> The total distance travelled by the object can be calculated by measuring the area between the graph and the baseline. This is called the area under the graph.

Work out the distance travelled by the trolley between:

Guided (a) 10 and 30 seconds **(2 marks)**

(30 – 10) × 20 = ... × 20 = ...

> The area of a triangle is: $\frac{1}{2}$ × base × height.

Guided (b) 0 and 10 seconds **(2 marks)**

0.5 × 10 × 20 = ...

(c) 30 and 70 seconds. **(2 marks)**

....

2 George is a vehicle technician. He is investigating the speed of a car over a 30 second period.

The car is stationary and then accelerates at 1.0 m/s^2 for 10 seconds. The car maintains a constant speed for an additional 20 seconds.

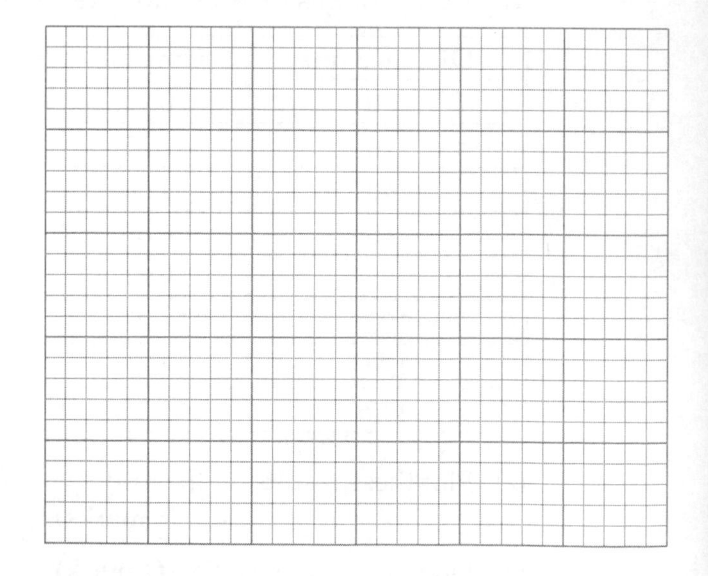

(a) Draw a graph on the grid to represent these data. **(6 marks)**

(b) Using the graph, work out the distance covered by the car between 10 and 20 seconds. **(2 marks)**

....

(c) Using the graph, work out the distance covered by the car between 0 and 10 seconds. **(3 marks)**

> Between 0 and 10 seconds the area under the graph is a triangle, so the distance travelled in this time will be the area of the triangle.

....

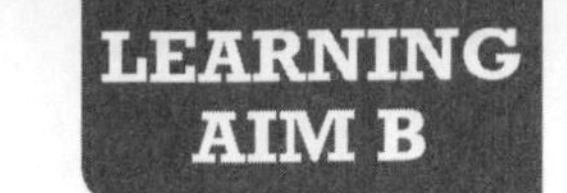

Why anomalous results occur

1 Complete the sentence by putting a cross in the box next to the correct answer.

The main reasons anomalous results occur is due to: **(1 mark)**

☐ **A** repeating measurements
☐ **B** having a large sample size
☐ **C** errors in measurements
☐ **D** inaccurate control variables.

2 A sports psychologist is carrying out an investigation into the effects of caffeine on a person's reaction times. They carry out a ruler drop experiment that tests reaction times with five males aged between 18 and 19 years.

For the first test, the subjects do not have any caffeinated drinks at all for 24 hours prior to the test. They carry out the reaction test three times and each score is recorded. For the second test, the subjects drink a coffee with two shots of espresso. After 10 minutes they repeat the reaction test a further three times.

The results are shown below.

Subject	Test results with no caffeine (s)	Test results after drinking caffeine (s)
1	0.48, 0.46, 0.46	0.42, 0.40, 0.41
2	0.52, 0.56, 0.59	0.52, 0.56, 0.53
3	0.42, 0.42, 0.43	0.39, 0.36, 0.33
4	0.41, 0.41, 0.42	0.36, 0.34, 0.33
5	0.35, 0.59, 0.59	0.55, 0.50, 0.49

(a) Circle the anomalous result in the table. **(1 mark)**

Guided

(b) Explain why this an anomalous result. **(3 marks)**

It is anomalous because it does not fit the pattern of the rest of the results and

....

....

....

(c) Give **two** possible reasons for this anomalous result. **(2 marks)**

....

....

There should be two reasons as there are two marks available for this question.

(d) Explain how this anomalous result should be dealt with when calculating the mean values for this person. **(2 marks)**

....

....

Positive correlation

1 Put a cross in the box next to the correct answer.

What type of relationship does the graph opposite represent? **(1 mark)**

☐ **A** positive correlation

☐ **B** no correlation

☐ **C** negative correlation

☐ **D** direct correlation.

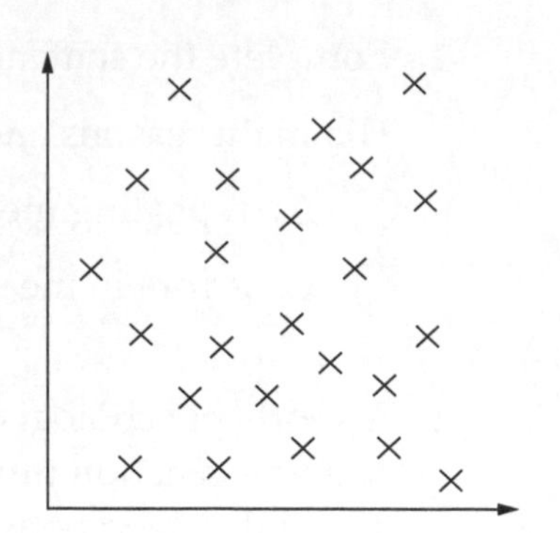

2 The graph below shows the data from an accelerating car.

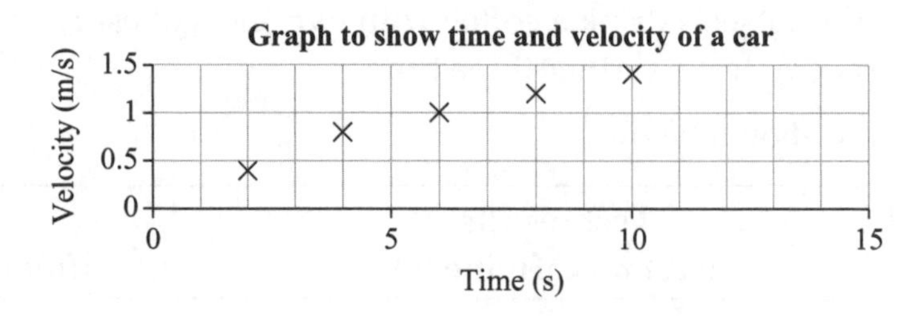

(a) Draw a best fit line on the graph. **(1 mark)**

Guided **(b)** Describe the correlation that is shown in this data. **(2 marks)**

A .... correlation is shown because when the time increases the

....

3 A health worker is investigating the effect of sugar intake on tooth decay. She has gathered data from four different world regions for the mean sugar intake per person and the population's mean number of decayed teeth.

The results are shown on the graph opposite and a line of best fit has been drawn.

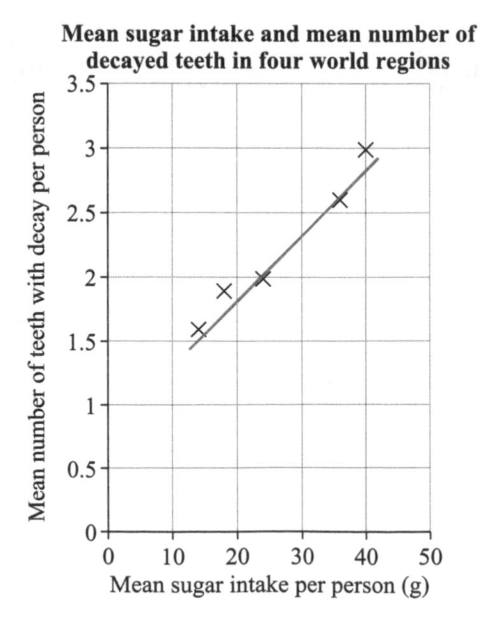

(a) Explain the type of relationship shown by the pattern on this graph. **(3 marks)**

....

....

....

Guided **(b)** Describe what conclusions could be drawn from this graph. **(2 marks)**

The more sugar a person eats ....

....

Negative correlation

1 Put a cross in the box next to the correct answer.

What type of relationship does the graph opposite represent? **(1 mark)**

☐ **A** positive correlation

☐ **B** no correlation

☐ **C** negative correlation

☐ **D** direct correlation.

In a positive correlation, as the independent variable increases, the dependent variable increases. In a negative correlation, as the independent variable increases, the dependent variable decreases.

2 Complete the sentence by putting a cross in the box next to the correct answer.

In a graph showing a negative correlation the best fit line: **(1 mark)**

☐ **A** slopes upwards ☐ **B** slopes downwards

☐ **C** passes through the origin ☐ **D** always has a steep gradient.

3 A nurse investigates the relationship between BMI and the amount of exercise people do each week.

The results of the investigation are shown below.

Number of exercise sessions per week (30 minutes per session)	0	1	2	3	4	5	6	7
Mean BMI	29	28	27	24	24	23	22	21

(a) Draw a graph of this data. **(6 marks)**

(b) Describe the correlation that is shown in the data. **(2 marks)**

....

....

....

....

....

....

Direct proportion

1 Put a cross in the box next to the correct answer.

Identify the type of relationship that the graph here represents. **(1 mark)**

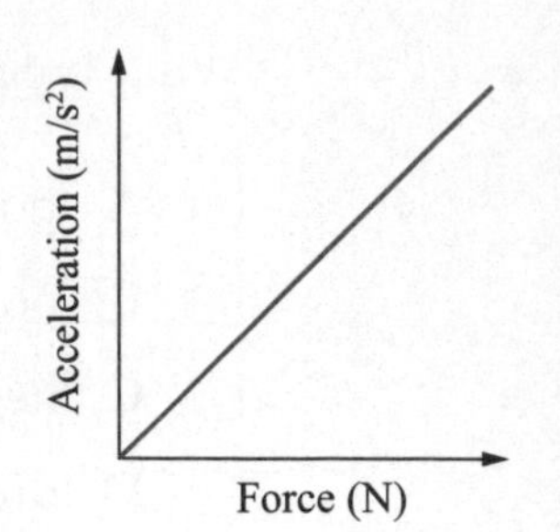

☐ **A** negative correlation ☐ **B** inversely proportional

☐ **C** no correlation ☐ **D** directly proportional.

A graph that is directly proportional has a straight line that follows a positive gradient and the line passes through the origin.

2 Hannah investigates how a spring is stretched when masses are suspended from it.

The results of her investigation are shown below.

Mass (g)	Length of spring (mm)			Mean length (mm)	Change in length (mm)
	Trial 1	Trial 2	Trial 3		
0	10	10	10	10	0
5	19	18	19		
10	27	26	26		
15	38	39	39		
20	50	49	49		
25	76	76	76		

(a) Complete the table by filling in the mean length and the change in length in mm. **(10 marks)**

Change in length needs to be measured against length with 0 g in each case.

(b) (i) Plot the data of the change in length against the mass. **(6 marks)**

(ii) Draw a best fit line. **(1 mark)**

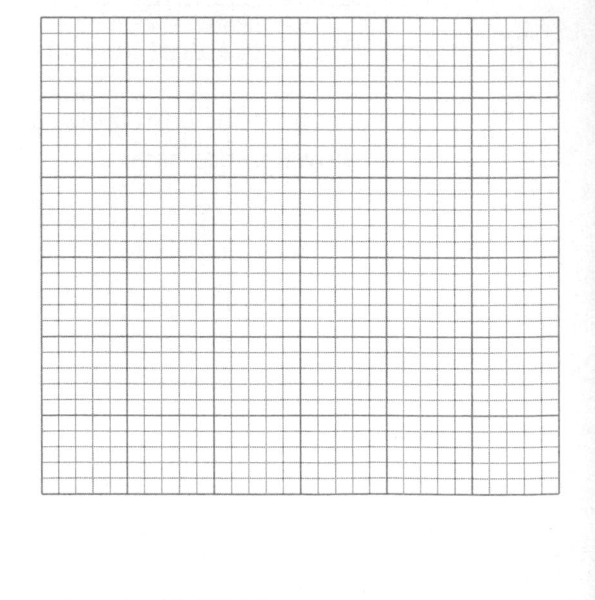

(c) Explain the relationship between the two variables. **(3 marks)**

The two variables are ....

....

because when one changes the other changes in the same way, by the same proportion.

....

....

Write one more sentence on how the best fit line indicates this relationship between the variables.

Inverse proportion

1 Complete the sentence by putting a cross in the box next to the correct answer.

A directly proportional graph shows the line of best fit passing through: **(1 mark)**

☐ **A** the y-axis

☐ **B** the x-axis

☐ **C** the origin

☐ **D** infinity.

2 A chemist performs an experiment to find out how the concentration of a substance affects the rate of a chemical reaction. She keeps the concentration of solution A constant and changes the concentration of solution B. Both substances are mixed and the rate of the reaction is recorded. The concentration of solution B and the time for the reaction to go to completion for each trial are recorded in the data table below and displayed in the graph.

Concentration of solution B (M)	0.005	0.010	0.015	0.020	0.025	0.030	0.035	0.040
Time to react (s)	20.0	12.0	7.0	4.5	3.2	3.1	3.0	2.9

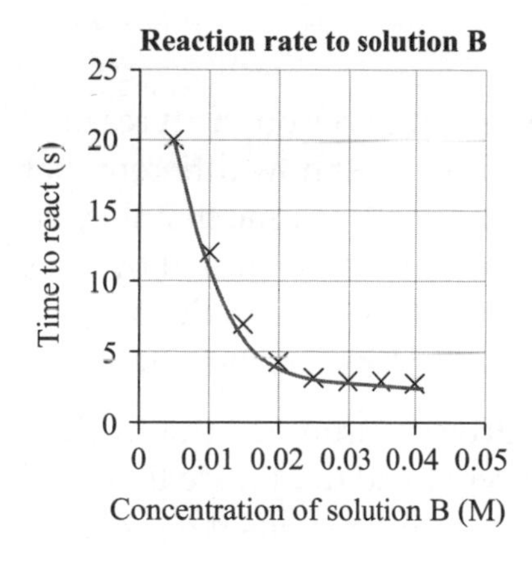

(a) State the type of relationship that this graph is showing. **(1 mark)**

....

Guided

(b) Describe how the concentration of solution B affects the time to react. **(2 marks)**

As the concentration of solution B ...., the time for the reaction to go to completion ....

....

Using evidence 1

1 Complete the sentence by putting a cross in the box next to the correct answer.

A conclusion is a decision that is made after the evidence collected has been: **(1 mark)**

☐ **A** analysed ☐ **B** extrapolated ☐ **C** discussed ☐ **D** repeated.

Guided

2 Alan carried out an experiment to compare the loss of water from the leaves of three different species of plant – A, B and C. He weighed 10 leaves of similar size from each species and hung them on a wire for 4 hours. He then weighed the leaves again and recorded the results.

Species	Mass of 10 leaves (g)	
	At start	After 4 hours
A	2.25	2.23
B	2.37	2.36
C	2.51	2.51

Use the information in the table to draw a conclusion about which plant species is the least efficient at retaining water in this investigation. **(2 marks)**

Species A lost 0.02 g of water, ...

.... Therefore species .... lost the most

water, so ....

3 Kevin works as a design engineer on the tyres of Formula 1 cars. He wants to test how different materials in tyres can increase the length of time that they can be used in races. Tyre durability can be tested by measuring the tread depth of a tyre. The tyre with the deepest tread depth after a set number of laps has the better durability.

He uses one car and drives it around a test track 10 times, then takes the tyre tread depth. He then replaces the tyres with another set and repeats the process with five different types of tyres. All of the tyres start with a tread depth of 1.75 mm.

The data from his test are shown in the table.

Tyre	Depth of tread (mm)
A	1.57
B	1.69
C	1.35
D	1.23
E	1.59

(a) Use these results to draw a conclusion about which tyre has the greatest durability. **(2 marks)**

....

....

....

(b) Can this conclusion be drawn from these data? **(2 marks)**

....

....

....

Check what the purpose of the investigation is to help you to decide if this conclusion can be drawn from these data.

Using evidence 2

Guided

1 Crude oil is a mixture of hydrocarbons that have different numbers of carbon atoms. A fractionating tower separates these mixtures based on the boiling points of the hydrocarbons.

Crude oil is heated to about 400 °C in a furnace to vaporise many of the hydrocarbons. The vaporised mixture is then pumped into a fractionating tower, where the temperature is highest at the bottom. The vaporised samples of hydrocarbons travel up the tower. They cool and condense at different levels and are collected on trays and removed.

Explain what conclusions can be drawn from this about the relationship between the boiling point of the hydrocarbons contained in the crude oil, and the number of carbon atoms in these molecules.

(4 marks)

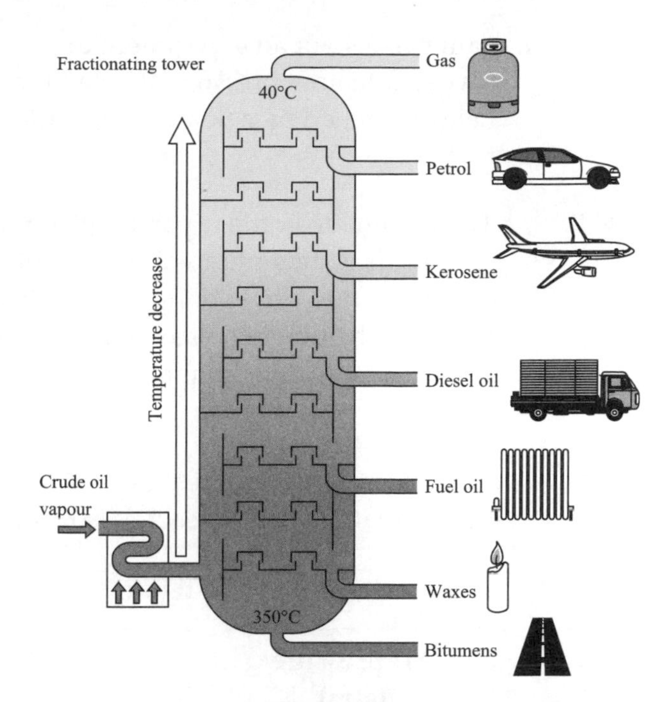

Fractional distillation of crude oil and the temperature range in which the different hydrocarbons condense.

Use your scientific knowledge to answer this question, based on work that you have carried out in Unit 5 Applications of chemical substances.

Molecules with longer hydrocarbon chains have higher boiling points. Bitumen is the hydrocarbon with the longest hydrocarbon chain as ....

....

....

....

....

Write one more sentence to explain why longer chained hydrocarbons have higher boiling points.

2 A forensic scientist has collected an unknown substance from a crime scene and is trying to find out what it is. She adds bromine water to the substance and the colour changes from orange to colourless.

The scientist concludes that the substance is an alcohol.

Is this conclusion correct? Explain your answer. **(2 marks)**

....

....

Had a go ☐ Nearly there ☐ Nailed it! ☐

Learning aim B sample questions

1 Sam carries out an experiment to investigate how pressure affected the height reached by a 'lemonade bottle rocket'.

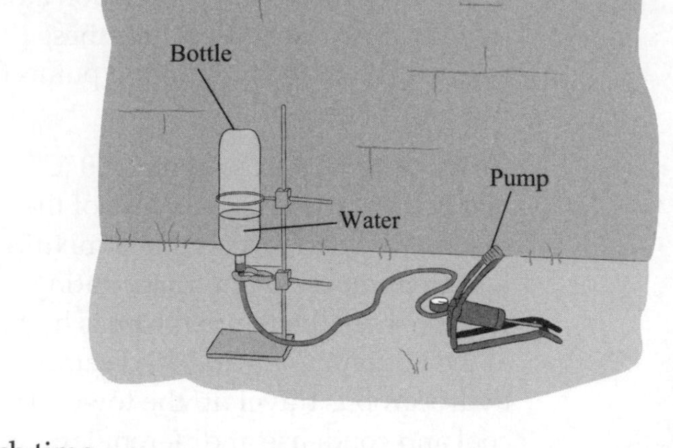

The lemonade bottle is partly filled with water. Air is then pumped in through a valve in the rubber bung. At a certain pressure the bung is forced out, which releases the water and makes the 'rocket' lift off.

The release pressure changes and the height that the rocket reaches is recorded each time.

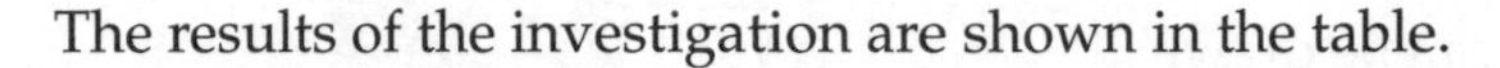

The results of the investigation are shown in the table.

Air pressure (atm)	Maximum height (cm)			Mean height (cm)
	Trial 1	Trial 2	Trial 3	
1.40	120	122	123	122
1.60	142	143	140	
1.80	162	163	161	
2.00	182	134	184	
2.20	202	200	204	202

(a) Complete the table with the mean height for each air pressure measurement. **(3 marks)**

(b) Explain the relationship between these variables. **(3 marks)**

....

....

....

....

Writing a conclusion 1

1 Complete the sentence by putting a cross in the box next to the correct answer.

A conclusion is: **(1 mark)**

☐ **A** an introduction to an investigation

☐ **B** a summary of an investigation

☐ **C** a prediction about what will happen in an investigation

☐ **D** a review of other studies related to the investigation.

2 Imran is a biomedical student. He is investigating how efficient ammonium chloride is as an ingredient in a chemical ice pack. The ice pack will contain water and ammonium chloride.

His hypothesis is: 'The more ammonium chloride that is added to water the greater the decrease in the temperature'.

Imran uses 100 cm^3 of water for each experiment and adds different masses of ammonium chloride. He takes the temperature before and after the ammonium chloride has been added and records the temperature difference for each amount. He repeats the test four times for each mass of ammonium chloride added. The results are shown below.

Mass of ammonium chloride (g)	Temperature difference (°C)				Mean temperature difference (°C)
	Test 1	Test 2	Test 3	Test 4	
10	–2.1	–2.0	–2.1	–2.1	–2.1
20	–3.2	–3.2	–3.1	–3.1	–3.2
30	–4.5	–4.5	–4.6	–4.5	–4.5
40	–5.7	–5.6	–5.7	–5.7	–5.7
50	–6.2	–6.3	–6.2	–6.2	–6.2
60	–7.1	–7.1	–7.0	–7.0	–7.1

(a) Can the hypothesis be accepted or rejected? **(1 mark)**

....

Guided

(b) Describe how the data support your answer to part **(a)**. **(2 marks)**

The results show a continued trend with increasing ....

....

(c) Explain, using scientific knowledge, why the temperature of the water decreases when ammonium chloride is added to it. **(3 marks)**

....

....

....

When ammonium chloride is added to water it produces an endothermic reaction.

Had a go ☐ Nearly there ☐ Nailed it! ☐

Writing a conclusion 2

Guided

1 Mark is carrying out an investigation to find out how air pressure affects the height reached by a rocket.
He states the hypothesis: 'The higher the pressure in the plastic bottle rocket, the greater the height it will reach'.

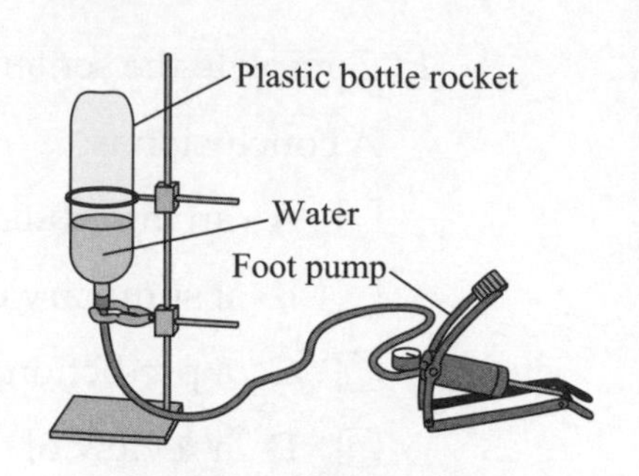

He sets up the experiment as shown in the diagram.

The plastic bottle is partly filled with water and air is pumped into the bottle using the foot pump. A rubber bung in the neck of the bottle includes a valve. When the pressure inside the bottle reaches a certain level, the pressure forces the bung out and releases the water, which causes the plastic bottle rocket to 'lift off'. The height of the bottle is recorded. He repeats the test three times for each release pressure.

A hypothesis states clearly what is expected to happen in an investigation, based on relevant scientific ideas.

The results are recorded in a table and displayed as a line graph.

Release pressure (atm)	Maximum height reached (m)			Mean height (m)
1.65	0.80	0.95	0.90	0.88
1.70	1.20	1.30	1.40	1.30
1.75	2.50	2.45	2.60	2.51
1.80	3.20	3.10	3.40	3.23

The effect of pressure on the height reached by a plastic bottle 'rocket'

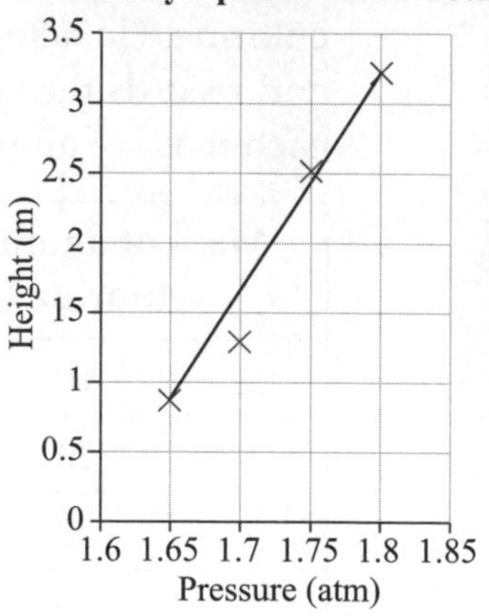

Write a conclusion for this experiment. **(6 marks)**

The hypothesis was: ....

This hypothesis can be .... The results show that for every increase in pressure the plastic bottle 'rocket' travelled to a greater height. The graph shows that

....

....

This increase in height is a result of ....

....

....

....

....

....

....

Inferences

1 A sports scientist studied the muscle cells of athletes and compared them to those of non-athletes. He found that the muscle cells of athletes have more mitochondria than muscle cells of non-athletes.

Mitochondria are the organelles where aerobic respiration (the release of energy from glucose or fat) takes place.

Complete the sentence by putting a cross in the box next to the correct answer.

Based on this investigation, it can be inferred that muscle cells in athletes: **(1 mark)**

☐ **A** contain more cytoplasm than non-athletes' muscle cells

☐ **B** reproduce less frequently than the muscle cells of non-athletes

☐ **C** have nuclei containing more DNA than nuclei in the muscle cells of non-athletes

☐ **D** have a greater demand for energy than the muscle cells of non-athletes.

An inference is a conclusion that is made based on limited information or evidence.

2 Sean is a microbiologist. He carried out an investigation into the factors that determine the effectiveness of penicillin in preventing bacterial growth.

He placed a small paper disc impregnated with penicillin on the surface of an agar plate over which the bacteria have been spread. The plate was then incubated for 3 days.

After this time an area free of bacterial growth could be seen around the disc of paper.

(a) Describe an inference that could be made from this investigation in relation to increasing the quantity of penicillin that the paper disc contained. **(2 marks)**

....

....

(b) State a hypothesis that could be developed based on this inference. **(1 mark)**

....

Guided

(c) Describe a further test that would need to be carried out to prove this hypothesis. **(2 marks)**

A further test would need to be carried out using different ....

....

to see if ....

Support for a hypothesis 1

1 Put a cross in the box next to the correct answer.

Identify which statement best describes a hypothesis. **(1 mark)**

☐ **A** A hypothesis is the process of making careful observations.

☐ **B** The conclusion drawn from the results of an investigation is part of a hypothesis.

☐ **C** A hypothesis is a basis for determining what data to collect when designing an investigation.

☐ **D** The facts collected from an investigation are written in the form of a hypothesis.

2 Scientists working for a drugs company are trying to find a substance that will prevent colds. The company have developed a chemical known as ColdK3. They state the hypothesis: 'The more ColdK3 a person takes the less chance they will catch a cold'.

They tested 20 000 volunteers, divided into four groups numbered 1, 2, 3 and 4. Each volunteer took a white pill every morning for 1 year. The contents of the pill taken by the members of each group are shown in the table below.

Group	Number of people in group	Contents of pill	Number of people who developed a cold
1	5000	5 g of sugar	20
2	5000	5 g of sugar, 1 g ColdK3	19
3	5000	5 g of sugar, 3 g ColdK3	20
4	5000	5 g sugar, 9 g ColdK3	18

Guided

(a) Do the data support the hypothesis: 'The more ColdK3 a person takes the less chance they will catch a cold'? Explain your answer. **(2 marks)**

No, the data do not support this hypothesis as ....

....

....

(b) Which statement is a valid inference based on the results? Put a cross in the box next to the correct answer. **(1 mark)**

☐ **A** Sugar reduced the number of colds.

☐ **B** Sugar increased the number of colds.

☐ **C** ColdK3 is always effective in the prevention of colds.

☐ **D** ColdK3 may not be effective in the prevention of colds.

(c) Which group was the control group in this investigation? **(1 mark)**

....

A **control** is a quantity that remains constant.

Support for a hypothesis 2

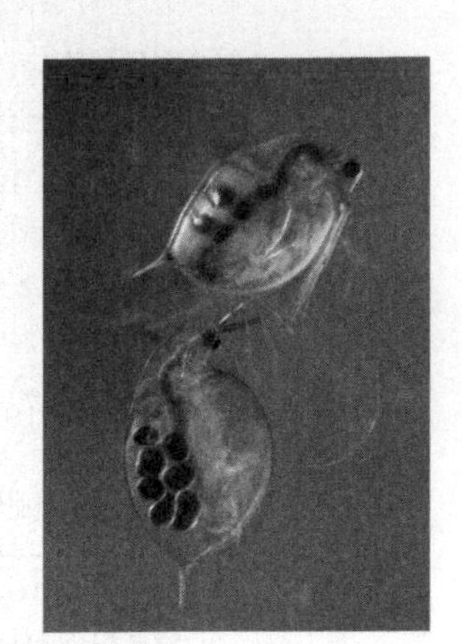

1 Marios is investigating the effect of water temperature on the heart rate of water fleas. His hypothesis is: 'As the temperature of water doubles, the heart rate of a water flea will double'.

He places water fleas in water at four different temperatures and takes their heart rates over a 1-minute period. He carries out each heart rate measurement four times and records the mean value.

His results are shown in the table below.

Water temperature (°C)	Mean water flea heart rate (bpm)
5	40
10	119
15	205
20	280

(a) Plot a graph of these data. **(6 marks)**

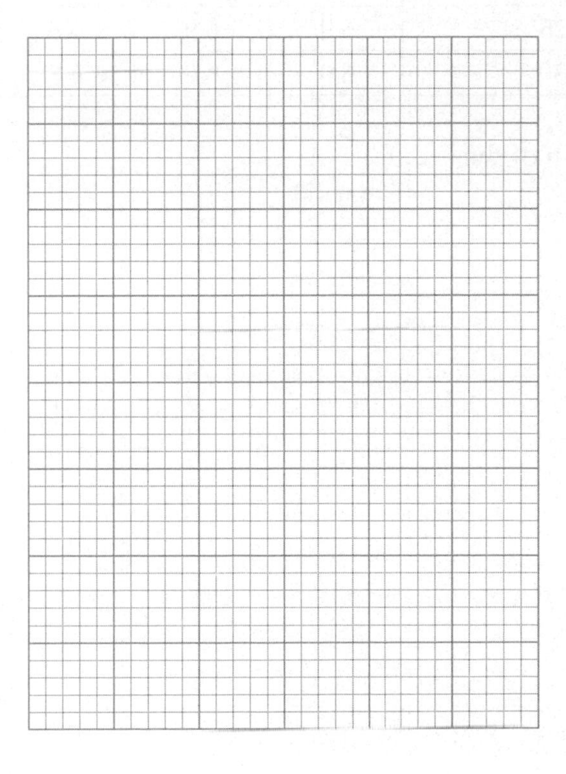

The independent variable is plotted on the *x*-axis and the dependent variable is plotted on the *y*-axis.

(b) Is the following hypothesis correct? 'As the temperature of water doubles, the heart rate of a water flea will double'. Explain your answer. **(3 marks)**

....

....

....

Guided

(c) State a qualitative hypothesis that would be supported by these results. **(1 mark)**

As the temperature of the water increases, ...

Support for a hypothesis 3

1 A biologist is investigating the effect of acid rain on pH levels in ponds. He takes 5 samples of water from ponds which are located at different distances away from city centres. His hypothesis is: 'The closer the pond is to the city centre, the lower the pH of the pond'.

The results of the investigation are shown below.

(a) Complete the table with the mean pH values for each pond. **(7 marks)**

Pond distance from city centre (km)	Sample pH reading					Mean pH
	1	2	3	4	5	
1	6.5	6.4	6.5	6.5	6.4	
5	6.5	6.5	6.4	6.5	6.5	
10	6.7	6.6	6.5	6.6	6.4	
15	6.8	6.7	6.6	6.8	6.8	
20	6.8	6.8	6.8	6.8	6.7	
25	6.9	6.9	6.8	6.8	6.9	
30	6.9	6.9	6.8	6.7	6.9	

(b) Plot a graph of these data. **(6 marks)**

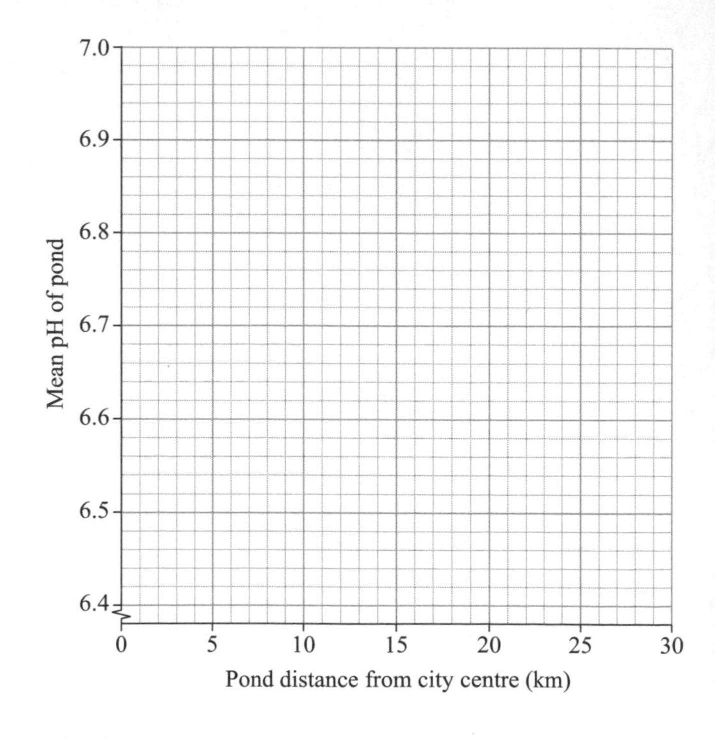

Guided

(c) Explain whether the evidence supports the hypothesis: 'The closer the pond is to the city centre, the lower the pH of the pond'. **(3 marks)**

The evidence supports this hypothesis up to a point, as the pH of the pond generally increases as the ponds get further away from the city centre. However, ...

....

....

Strength of evidence

1 Complete the sentence by putting a cross in the box next to the correct answer.

If the evidence does not allow you to be certain about the conclusion, you should state that the hypothesis is: **(1 mark)**

☐ **A** supported by the evidence

☐ **B** strongly supported by the evidence

☐ **C** not supported by the evidence

☐ **D** weakly supported by the evidence.

2 Anya is investigating the extension of a spring when different weights are added to it. Her hypothesis is that the length of the spring will increase as more weight is added.

The results of her investigation are shown in the table and the points plotted on a graph.

Attached weight (N)	Total length of spring (m)
1	0.37
2	0.42
3	0.51
4	0.59
5	0.64

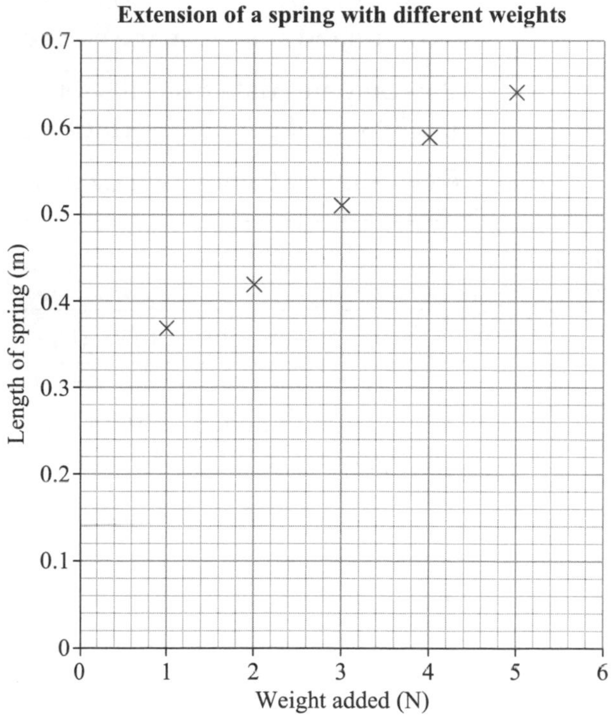

(a) On the graph, draw the line of best fit. **(1 mark)**

To draw a line of best fit, exclude any anomalous data and draw the line so that about half the points are on either side of it.

Guided

(b) What sort of relationship is shown between the independent and the dependent variables? **(2 marks)**

There is strong .... correlation showing that as more force is added to the spring, the ....

(c) Is the hypothesis: 'The length of the spring will increase as more force is added' strongly supported by this evidence? Explain your answer. **(3 marks)**

....

....

....

Had a go ☐ Nearly there ☐ Nailed it! ☐

Sufficient data 1

1 Complete the sentence by putting a cross in the box next to the correct answer.

In an investigation how many times should repeat measurements be taken to check that the results are repeatable? At least: **(1 mark)**

☐ **A** 2 times

☐ **B** 3 times

☐ **C** 4 times

☐ **D** 5 times.

2 A health visitor states the hypothesis: 'Second-hand smoke has a negative effect on the growth of a baby in the womb'. She carries out an investigation into the effect of second-hand smoke on the mass of babies at birth.

She compares data from 100 sets of parents consisting of 50 sets of parents that do not smoke and 50 sets of parents in which the mother does not smoke but the father does smoke.

The results of the investigation are shown below.

	Both parents non-smokers	Mother non-smoker, father smoker
Mean mass of baby at birth (kg)	3.2	3.0

Guided

(a) Examine the data that has been has been collected and recorded. Explain if the hypothesis: 'Second-hand smoke has a negative effect on the growth of a baby in the womb' can be accepted. **(6 marks)**

The results suggest that second-hand smoke ....

....

....

However the masses at birth are very close together – there is only 0.2 kg difference between the two values. This means that ....

....

....

(b) Explain if the data collected are reliable. **(2 marks)**

....

....

An investigation should include repeating measurements at least three times to provide reliable data.

Sufficient data 2

1 Complete the sentence by putting a cross in the box next to the correct answer.

To be sure of a relationship on a graph, there need to be at least: **(1 mark)**

☐ **A** 3 data points ☐ **B** 4 data points ☐ **C** 5 data points ☐ **D** 6 data points.

2 Complete the sentence by putting a cross in the box next to the correct answer.

If another person carries out your investigation and they obtain similar results, the data are: **(1 mark)**

☐ **A** reliable ☐ **B** correlated ☐ **C** anomalous ☐ **D** related.

3 Matt is a technician at an artificial ski slope. He plans to construct a ski jump and wants to investigate how the height of the jump affects the distance travelled by a skier after he lands.

He sets up a model using a ball bearing and science lab equipment as shown below.

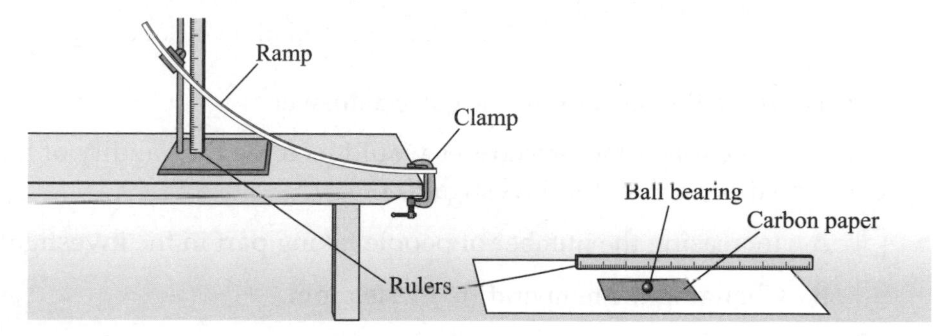

The ball bearing represents the skier. It rolls down the ramp and lands on a sheet of carbon paper, placed on a piece of white paper taped to the floor. This makes a mark on the white paper so that the horizontal distance that the ball bearing travels can be measured. The height of the ramp can be changed.

The results of the investigation are recorded in the table below and displayed as a graph.

Height of ramp (cm)	Distance travelled (cm)
20	85.5
30	84.5
40	84

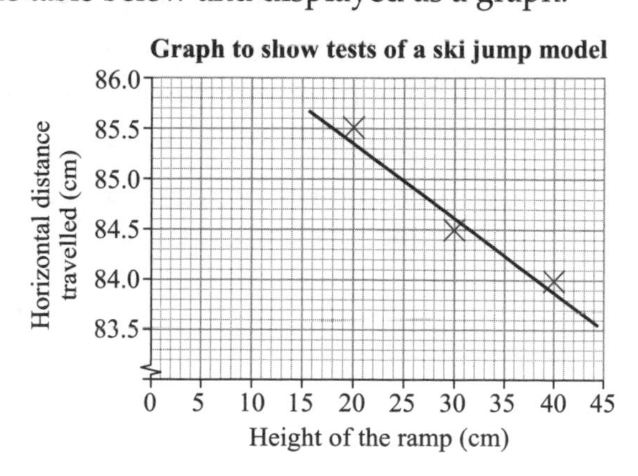

Matt states that there is a negative correlation between the height of the ramp and the horizontal distance travelled.

Can this conclusion be drawn from this amount of data? Explain your answer. **(3 marks)**

....

....

....

Had a go ☐ Nearly there ☐ Nailed it! ☐

Validity of a conclusion

1 Harry wanted to investigate the effect of exercise on heart rate. Five 16 year old males take part in 1 minute of skipping. He measured their heart rate before the 1 minute of skipping and then immediately after they stopped skipping. The results are shown in the table below.

Person	Pulse rate before skipping (bpm)	Pulse rate after skipping (bpm)
1	70	98
2	74	102
3	82	118
4	61	82
5	77	92

(a) State a valid conclusion that can be drawn from these results. **(1 mark)**

....

(b) Put a cross in the box next to the correct answer.

Which of the following procedures would increase the validity of the conclusion that could be drawn from this investigation? **(1 mark)**

☐ **A** Increasing the number of people taking part in the investigation.

☐ **B** Changing the temperature in the room.

☐ **C** Decreasing the number of students participating in the activity.

☐ **D** Increasing the time spent skipping.

Guided

2 A scientist carried out an investigation to test the hypothesis that marigold seeds exposed to acid rain will not grow at the same rate as marigold seeds exposed to normal rain.

The scientist set up two groups, each containing 20 marigold seeds. He watered group A with normal rain water with a pH of 5.6 and group B with acidic water with a pH of 4.0. All other conditions were kept the same.

After 21 days, he recorded the growth of the plants. He then recorded the mean shown in the table below.

	Group A: normal rain water (pH 5.6)	Group B: acid rain water (pH 4.0)
Mean plant growth (cm)	3.4	2.6

The scientist concluded that the acid rain slows down the rate of photosynthesis of plants. Is this is a valid conclusion? Explain your answer. **(3 marks)**

This is not a valid conclusion because ....

....

Remember photosynthesis is the process where plants use sunlight and carbon dioxide to make glucose.

Evaluating an investigation 1

1 Annmarie works for a manufacturer producing suncreams that block the harmful UV rays emitted from the Sun. Her hypothesis is that the UV rays will be at their highest levels at 2 pm in the UK. If her hypothesis is correct she can tell people when they need to be more careful about wearing suncream and the SPF factor that they should wear.

SPF suncream is used to block the UV rays from damaging the skin. It is a number on a scale for rating the degree of protection provided by sunscreens. SPF stands for Sun Protection Factor.

UV rays come from the Sun and can burn the skin.

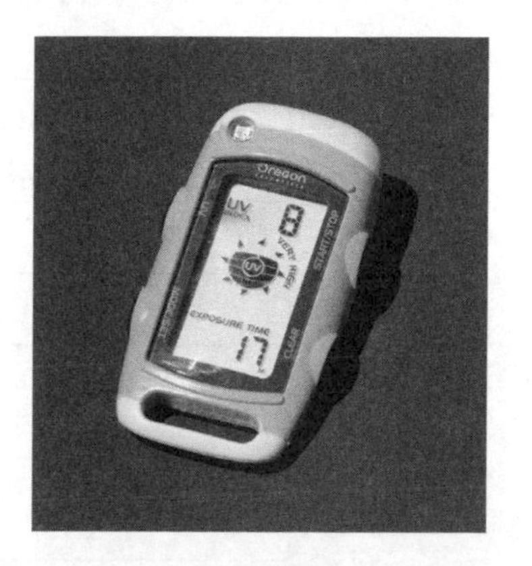

She uses a UV monitor for this investigation which records from 0 for no UV rays to 25 for the highest UV reading.

She finds an area outside that receives direct sunlight all day long and selects that area to take the UV readings. Over a 4 day period during the month of July, she takes and records UV readings every 2 hours between 10 am and 4 pm.

	UV index				Mean UV index
Day	Mon	Tues	Wed	Thurs	
10 am	5	7	3	8	5.85
12 pm	15	18	6	19	14.5
2 pm	18	19	6	21	16.0
4 pm	14	16	5	15	12.5

A bigger range of results produced in an investigation will help find a pattern or trend. Independent variable values should be selected that are evenly spaced, if possible, and at least six different values should be included.

From this investigation she concludes that UV rating in Britain is at its highest at 2 pm so she can accept her hypothesis. So that is the time of day when it is most important to wear suncream.

(a) Were enough readings taken each day during this investigation to accept the hypothesis that UV rays are at their highest levels at 2 pm? Explain your answer. **(2 marks)**

....

....

Guided

(b) On Wednesday the UV ratings are all very low in comparison to the other days.

Describe **one** factor that could have contributed to these low UV ratings. **(2 marks)**

If it had been a cloudy day the UV ratings would have been .... because the clouds ....

(c) Does the 4 day period for this investigation provide enough evidence to conclude that the UV rays in Britain are at their highest levels at 2 pm? Explain your answer. **(2 marks)**

....

....

Had a go ☐ Nearly there ☐ Nailed it! ☐

Evaluating an investigation 2

1 Luca is trying to work out how much calcium oxide would be best to use in a hand warmer. He carries out an investigation using different masses of calcium oxide and adds this to 30 cm^3 of water.

He records the temperature of the water, measures out different masses of calcium oxide on digital scales and mixes them with the water. He takes the temperature of the water and calcium oxide at intervals until they have reached their highest temperature.

The results of the investigation are shown in the table below and displayed in a graph.

Mass of calcium oxide (g)	Temperature rise (°C)
3	4
6	6
9	7
12	12
15	11

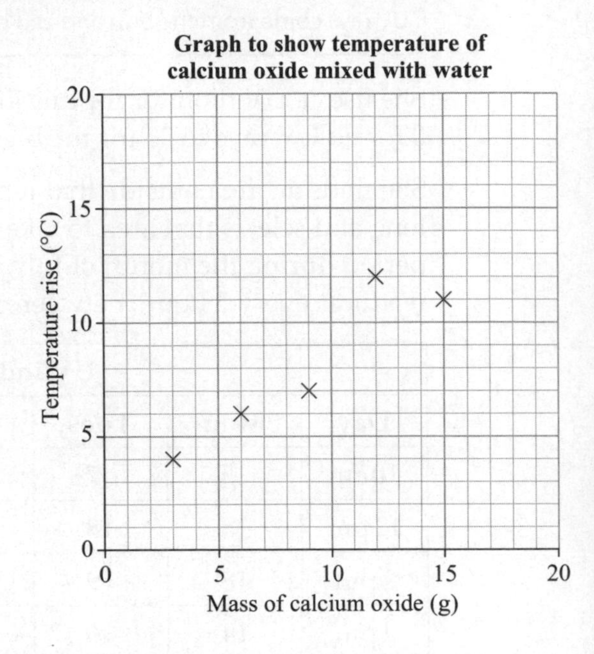

He concludes that the more calcium oxide is added to water, the higher the temperature rise.

(a) Draw a line of best fit on the graph. **(1 mark)**

(b) Circle the anomalous result in these data. **(1 mark)**

(c) Explain if the data collected are reliable. **(2 marks)**

....

....

Guided

(d) Explain **one** weakness in the method used in this investigation. **(2 marks)**

Only one reading for each mass of calcium oxide was taken. It would have been better to

....

....

Anomalous data is usually due to errors in the experimental process so look at the method to consider what may have gone wrong.

Improving an investigation 1

1 A forensic scientist is investigating the effect of height on the number of spines shown on blood droplets when they hit the ground. Her hypothesis is: 'The greater the distance the droplet has to fall to the ground, the more spines form around the edges of the droplet'.

She uses a pipette and drops out a measured quantity from different heights. She then counts the number of spines formed around the edge of each droplet.

She repeats the experiment five times for each height. The results are recorded in the table.

Height (mm)	Number of spines					
	Test 1	Test 2	Test 3	Test 4	Test 5	Mean
10	5	6	4	5	6	5
20	6	7	6	8	8	7
30	8	9	9	8	9	9
40	10	9	10	8	9	9
50	12	13	13	10	12	12
60	12	13	14	12	11	12
70	14	15	14	12	15	14
80	13	14	16	15	14	14

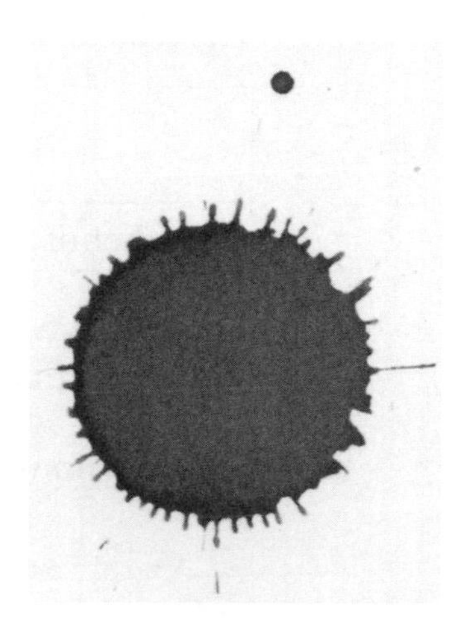

(a) Plot these data on a graph.

(6 marks)

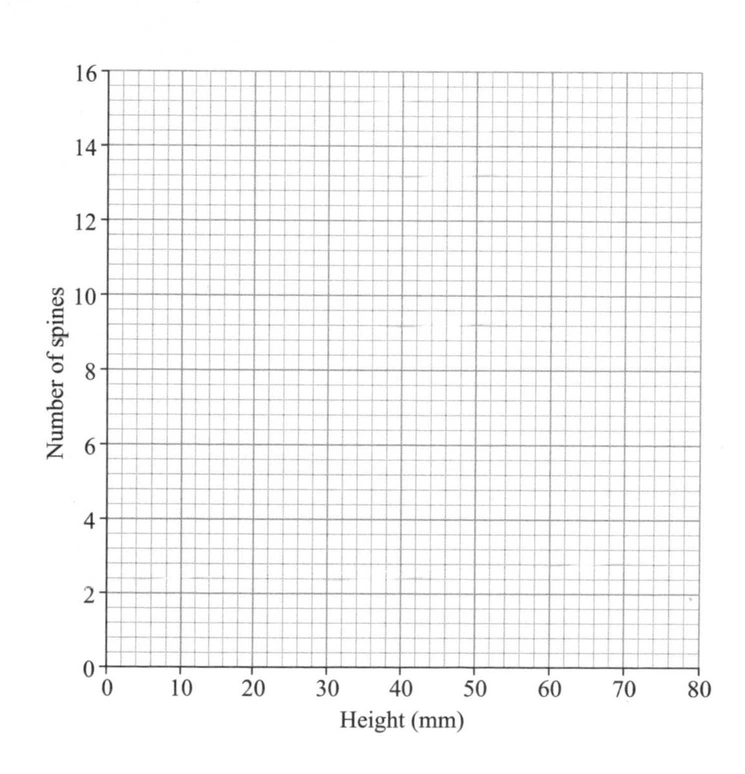

Guided

(b) Explain **one** way that this investigation could be improved. **(3 marks)**

They could increase the range of the independent variable as ....

....

Think about the average height of a person compared to the range of the independent variable used.

Improving an investigation 2

1 A cosmetics company has produced a new nanochemical that they add to mascaras. The nanochemical is designed to increase the length of a person's eyelashes.

The new mascara (Mascara A) is given to 100 women aged between 25 and 30 to test. The women also test a mascara that does not contain the nanochemical (Mascara B). The volunteers were asked to state which mascara made their eyelashes look longer or if there was no difference between the two.

The results of the investigation are shown in the table below.

	Mascara A made eyelashes look longer	**Mascara B made eyelashes look longer**	**No difference**
Number of volunteers	65	25	10

The head scientist looked at the data and said that the investigation was not very reliable as the results that were obtained were subjective.

(a) Explain what is meant by 'subjective'. **(2 marks)**

....

....

Guided

(b) Describe a method that could be used to improve the investigation so that the data obtained are not subjective. **(2 marks)**

The mascara is designed to increase the length of a person's eyelashes. Therefore instead of asking ....

....

they could ....

....

(c) The head scientist has suggested that as well as taking non-subjective measurements, the investigation could be improved if more volunteers were tested.

Explain why increasing the sample size will improve the investigation. **(2 marks)**

....

....

Extending an investigation

1 Complete the sentence by putting a cross in the box next to the correct answer.

One way to extend an investigation is to: **(1 mark)**

☐ **A** test a related hypothesis

☐ **B** repeat the investigation

☐ **C** test the same sample again

☐ **D** increase the range of measurements.

2 A sports scientist is investigating the effect of sports drinks on endurance exercise. The sports scientist's hypothesis is: 'By drinking a sports drink, the sugar in the sports drink will supply energy to the exercising muscles, which will allow a person to exercise for longer as their stored glycogen supplies will last longer'.

A sports drink that is 9% glucose is used in the investigation.

The subjects run on a treadmill at a set speed for as long as they are able to until exhaustion.

Two tests are carried out on 10 male subjects. In test 1, the subject must consume 100 cm^3 of water during each 30 minute period throughout the exercise period. Two weeks later, in test 2, the sports drink is given instead of water.

The results are shown in the table on the right.

The sports scientist accepts his hypothesis that drinking a 9% glucose sports drink will improve endurance exercise performance.

Subject number	Time spent exercising with water (min)	Time spent exercising with 9% glucose sports drink (min)
1	65	72
2	48	49
3	82	98
4	52	56
5	46	45
6	89	99
7	120	128
8	86	90
9	117	119
10	79	82

(a) Suggest how the independent variable could be changed to extend this investigation. **(2 marks)**

The independent variable is the one that the scientist controls.

....

(b) Suggest how the conditions of the investigation could be changed to extend the investigation. **(2 marks)**

....

(c) Suggest how the sample being tested could be changed to extend the investigation. **(2 marks)**

....

....

Learning aim C sample questions

1 A scientist working for a pharmaceutical company is testing out a new indigestion tablet. They have 5 different tablets labelled A, B, C, D and E.

The scientist uses a pH meter to see how the pH of dilute sulfuric acid changes when she adds one dose of each indigestion tablet to the acid.

The results of the investigation are shown below.

Tablet	pH of sulfuric acid after the indigestion tablet has been added
A	2.9
B	3.2
C	2.9
D	3.8
E	4.1

From this investigation the scientist concludes that tablet E is the most suitable for use as an indigestion remedy.

(a) Explain how the scientist can control **three** different variables to make this a fair test.
(3 marks)

....

....

....

....

(b) Explain how this investigation could be improved. **(3 marks)**

....

....

....

....

Acid that is produced by the stomach and causes indigestion is hydrochloric acid.

Answers

Learning aim A

1. Scientific equipment 1

1 B
2 It is not clean. It may contaminate the experiment which could produce inaccurate results.
3 A – Bunsen burner
B – heatproof mat
C – rubber tubing
4 A beaker can be heated, a flask can be heated and a boiling tube can be heated.
Other answers include a test tube and a crucible.

2. Scientific equipment 2

1 B
2 **(a)** The Bunsen burner will heat up the test tube and the water inside it which will then become too hot for a person to hold without burning themselves. The tongs are used so that the person does not have to touch the test tube so that they do not get burnt.
(b) He could use a clamp and a clamp stand to secure the test tube over the Bunsen burner flame.
(c) Test tubes should be stored in a test tube rack to keep them upright to prevent their contents from spilling out.
(d) A stirring rod could be used to mix the substance.

3. Chemistry equipment

1 A measuring cylinder.
2 She could use a pH meter or universal indictor solution.
3 **(a)** Filter paper and a filter funnel are used to separate a solid – the sand, from a liquid – water, in the mixture. The filter paper is folded into quarters and placed in the funnel. The mixture is then poured onto the filter paper and the water filters through, leaving the sand behind on the filter paper.
(b) A digital balance could be used to measure the mass.

4. Physics equipment

1 C
2 **(a)** Ⓐ ammeter
Ⓥ voltmeter
⊗ filament bulb
(b) Ammeter – measures the current in a circuit.
Voltmeter – measures the voltage across a circuit component.
Filament bulb – to show that the circuit is working/ investigate its resistance.
(c) A resistor.
3 He could use light gates set up at each of the distance intervals that he wants to measure. These work by recording the time at which the athlete passes through each of the light gates, which are measured over set distances along the 50 m running track.

5. Biology equipment

1 **(a)** C
(b) She will need a glass microscopic slide to place the object on and a glass cover slip to place over the sample to hold it in place. This will prevent the sample from coming into contact with the microscope lens.
2 He will need a strain of bacteria, five Petri dishes to grow the colonies safely in the lab and also agar to provide nutrients for the bacteria that are placed in the Petri dishes.
3 To measure the children's mass a set of bathroom scales can be used as this measures mass in kilograms. The scales also need to give the mass to 1/10 of a kg. Height can be measured using a tape measure which measures at least 2 m with cm markings as no 8–9 year old children are likely to be taller than this height. The person stands without their shoes on against a wall and a mark is made in line with the top of their head, the tape measure is then used to measure their height up to the mark drawn on the wall.

6. Risks and management of risks

1 B
2 One hazard is the Bunsen burner. Another hazard is the corrosive substance.
One risk is her loose hair. The other is that she is not wearing safety goggles.
3 A control measure is something that is put in place to reduce the identified risks associated with an experiment.
4 He should carry this experiment out in a fume cupboard because chlorine gas is toxic and if he breathes it in it will damage his lungs. If the experiment is carried out in a fume cupboard the toxic fumes are removed and cannot be breathed in. He should also wear safety goggles, a lab coat and gloves.

7. Hazardous substances and control measures

1 A
2 **(a)** It is flammable so it should be stored in flame-resistant cupboards.
(b) It is corrosive so she must avoid contact with the skin as it attacks and destroys living tissue.
(c) This substance is not corrosive but is harmful if inhaled or will cause irritation to the skin, eyes or inside the body.
(d) This substance could cause immediate or delayed harm to the environment.
3 Corrosive substances destroy living tissues so safety goggles should be worn to protect the eyes by preventing the substance from being splashed into their eyes.
4 The substance should be used in a fume cupboard so that the harmful fumes are not breathed in. A person should also wear eye protection to protect the eyes from splashes and gloves to protect the skin when using these types of substances.

8. Protective clothing

1 C
2 Item A is safety goggles. They are worn to protect the eyes from a range of hazards including splashes from corrosive or hot liquids.
Item B is a lab coat. It is worn to protect skin and clothes from splashes, spills and flames. It can be removed very quickly if needed.
Item C is gloves. They are worn to protect the hands from harmful substances or to prevent the spread of infection if bacterial cultures are being grown. Gloves can also be worn to protect the hands from heat or cold.

9. Handling microorganisms

1 B
2 She should wear protective gloves.
She should wash her hands with an antibacterial soap after handling the equipment, and before eating or drinking anything.
3 1 Wipe down the surface with an antibacterial spray after use.
2 Wash hands afterwards with antibacterial spray.
3 Do not eat or drink while carrying out the experiment.

10. Microorganisms and wildlife

1 B
2 The lid is taped to the dish to stop microorganisms from the air getting in.
The Petri dish is incubated at 25 °C. Temperatures higher than this encourage the rapid growth of bacteria and that might include pathogens.
3 **(a)** The children are handling worms that have been in the soil so they may have been in contact with dangerous microorganisms which could be harmful to the children.
(b) The children should wash their hands with antibacterial soap to remove any harmful microorganisms.

11. Dependent and independent variables

1 B
2 **(a)** The concentration of HCl.
(b) Rate of reaction.
3 **(a)** The independent variable is the glucose concentration

in the sports drink because the quantities of glucose in each are known by the sports scientist.

(b) The dependent variable is the blood glucose concentration because this is what is measured by the sports scientist. The changes that are made to the independent variable will have a direct effect on blood glucose concentrations of the athlete.

12. Control variables

1 B

2 (a) Two of the control variables are the quantity of compost and the quantity of water that each plant receives. Another control variable is the fact that both groups are in the same greenhouse as it means they are receiving the same amount of light/heat.

(b) The control variables are used so that if there is any differences in the firmness of the tomatoes when they are ripe, the scientist can be relatively sure that this is due to the fact that they are from genetically engineered plants rather than other environmental factors.

3 The mass of the water for each experiment would have to be the same – if more water was used for one experiment than another then it would take more energy to change the temperature of the water so the results would not be accurate.
The amount of each fraction of crude oil must also be kept the same for each experiment as if there was a greater mass of one type of crude oil fraction, it would burn for longer. This may increase the temperature of the water to a greater level and therefore give inaccurate results.

13. Measurements

1

Size	Number of metres
1 kilometre	1000
1 metre	1
100 kilometres	100 000
1 centimetre	0.01
1 millimetre	0.001
10 millimetres	0.01
1 nanometre	0.000 000 001
10 centimetres	0.1

2 (a) 2×10^4

(b) 1×10^{-3}

(c) 1.5×10^8

3 6×10^{-5} m

14. Units of measurement

1 B
2 C
3 B
4 grams per cubic centimetre
5 watts or kilowatts
6 ohms
7 B
8 A
9 B
10 (a) This animal should be weighed in grams as it is not very heavy.
(b) This animal should be weighed in kg as it is quite heavy.

15. Accurate and precise measurements 1

1 B

2 (a) This set of measurements is measuring the same thing and all the measurements are close together. 0.02 m/s is the largest difference between the highest and lowest values. Precise measurements are those that, when repeated, are consistent and have a limited range, which means that these measurements are precise.

(b) Accurate measurements are ones that are close to the correct value. As the mass of the trolley was not accurate, the results that have been recorded for this part of the investigation are not accurate.

(c) It is possible for a set of results to be precise but not accurate because this means most of the measurements taken are close together (precise) even if, for some reason they are not actually correct (accurate).

16. Accurate and precise measurements 2

1 (a) (i) 66 bpm
(ii) 66 bpm

(b) The results are accurate as the manual radial pulse readings are very close or the same as the heart rate monitor results which confirms that both sets of data should be accurate. The range of results are also very close together with only 2 bpm difference in the highest and lowest readings which shows that the readings are also precise.

17. Range and number of measurements

1 (a) It does not follow the pattern of the rest of the results where there is only –0.1°C of a temperature difference. Here there is a –0.7°C difference compared to the rest of the results for this mass of ammonium chloride.

(b) Yes, the results show good repeatability as each time the experiment was repeated the spread of measurements for each data set is very small, only 0.1°C difference in most instances.

(c) Six different masses of ammonium chloride have been used which is a sufficient number of tests as these data can be plotted on a graph. A line of best fit can be drawn so that a trend can be seen.

(d) The reaction is an endothermic reaction, which means that heat energy is taken in to form bonds. This means that the contents of the packet get colder.

18. Writing a method

1
- a list of the equipment and substances needed
- how much of each substance is needed
- what measurements you will take and how many times you need to measure them.

2 (a) A ruler or measuring tape.

(b) Yes, the quantities of fertiliser, potting soil and water have all been provided in the method.

(c) Having all the plants with the same quantity of potting soil and giving both plants the same quantity of fertiliser are two control variables used in this method.

19. Hypotheses

1 (a) Any one from:
Rose plants grow faster at 20°C than at 15°C.
Temperature affects plant growth.
Rose plants produce more flowers at higher temperatures.

(b) Answer depends on hypotheses stated in part (a), e.g.
height
mass
number of leaves
number of flowers.

2 Any one from:
Hand grip strength will decrease with each consecutive trial.
As the number of trials increases, the hand grip strength decreases.

3 The more vitamin C a person with a cold consumes the less time they will suffer from the symptoms of a cold.

20. Qualitative and quantitative hypotheses

1 Quantitative

2 A qualitative hypothesis describes what is expected to happen but does not include a numerical prediction. A quantitative hypothesis gives an expected value, percentage or factor change.

3 Qualitative

4 If females who are 30 kg or more overweight consume 1500 kcalories per day they will lose 1 kg of body mass per week.

21. Planning an investigation

1 Aim: The purpose of this experiment is to determine whether exercise improves the rate at which maths problems are solved.
Hypothesis: I think that exercise will improve the rate at which maths problems are solved.
Method: There will be two groups in the experiment – groups A and B.
Group A is the experimental group and they will take part in vigorous exercise for at least 15 minutes running on a treadmill. Group B is the control group and will not take part in any exercise.
Both groups will be given the same mathematical problems and be timed to work out the speed with which they complete the problems/or the number of maths problems solved in a certain amount of time.
If group A solves the maths problems more rapidly than the control group (or solves more problems over a given time period) then this will support the claim by the mathematician.

22. Learning aim A sample questions

1 **(a)** Independent variable is the concentration of antibiotic. Dependent variable is the diameter of the exclusion zone.
(b) The temperature of the incubator has to be kept the same in the repeat experiment as in the first experiment. This is so that the bacteria are given the same temperature to grow in as different temperatures can affect the rate of growth of bacteria.
The same agar nutrient should be used in each plate because different nutrients might affect the rate of growth of the bacteria feeding on the nutrient in each plate.

Learning aim B

23. Tables of data

1 C

2

Tyre	Tread depth (mm)
A	1.57
B	1.69
C	1.35
D	1.23
E	1.59

3 **(a)** Speed
(b) First column.

4

Time (min)	Temperature (°C)
0	16
1	18
2	24
3	28
4	33
5	33
6	33

24. Identifying anomalous results from tables

1 D

2 **(a)**

Sugar solution concentration (g/l)	Mean length after 24 hours (mm)
0	51
5	43
8	43
10	41

(b) 51 mm in 8 g/l sugar solution.
(c) This is an anomalous result because it falls outside the range of the other results in the group at 8 g/l sugar solution. The other results in this sugar solution range from 42 to 44 whereas this reading is 51 mm. The range in the other solutions is around 2 mm difference so this large difference does not fit the pattern of the other results.

25. Identifying anomalous results from graphs

1 **(a)**

Mass of ammonium chloride (g)	3	6	9	12	15
Temperature rise (°C)	4	6	7	12	11

(b) and (c)

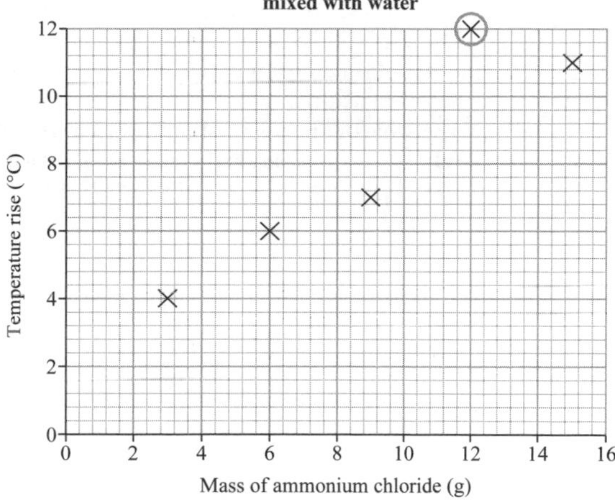

The reading for 12 °C at 12 g ammonium chloride is anomalous.
(d) This is an anomalous result because it falls outside the pattern of the other results. The pattern is a steady increase in temperature as more ammonium chloride is added. At 12 g of ammonium chloride added the temperature rises steeply, which is not in line with the pattern shown by the rest of the results.

26. Excluding anomalous results

1 D

2 **(a)** Sample 11 – 3.68%
(b) It should be ignored and left out.
(c) 48.98/14 = 3.50%

27. Calculations from tabulated data

1 **(a)**

Person	Test results after heating hands in a water bath at 40 °C (seconds)	Mean test result after heating (seconds)	Test results after cooling hands in a water bath at 5 °C (seconds)	Mean test result after cooling (seconds)
1	0.42, 0.40, 0.41	0.41	0.48, 0.46, 0.46	0.47
2	0.52, 0.56, 0.53	0.54	0.52, 0.56, 0.59	0.56
3	0.39, 0.36, 0.33	0.36	0.42, 0.42, 0.43	0.42
4	0.36, 0.34, 0.33	0.34	0.41, 0.41, 0.42	0.41
5	0.55, 0.50, 0.49	0.51	0.60, 0.59, 0.59	0.59

(b) (i) 0.43 seconds
(ii) 0.49 seconds

28. Calculations from tabulated data – using equations

1 (a) $6 + 4 + 5 + 4 + 8 + 3 = 30$
Number of rose plants = 6
Total number of flowers divided by number of plants = $30 \div 6 = 5$
(b) $50 \div 6 = 8$

2 $0.058 \div 0.6 = 0.097$ m/s

29. Significant figures

1 A

2 (a) 16000
(b) 15600
(c) 15650

3

Person	Mass (kg)	Mass to two significant figures (kg)
a	62.34	62
2	79.81	80
3	75.50	76
4	68.29	68
5	55.61	56

30. Bar charts 1

1 A

2 The chart should have a title that provides a summary of the chart content.
The names of the variables should be written on each axis.
The units should be written on each axis if appropriate.
The bars should be drawn with equal widths.
A gap should be left between each bar to make the chart easier to read.
The scale on the y-axis should start at zero/0.

3 (a)

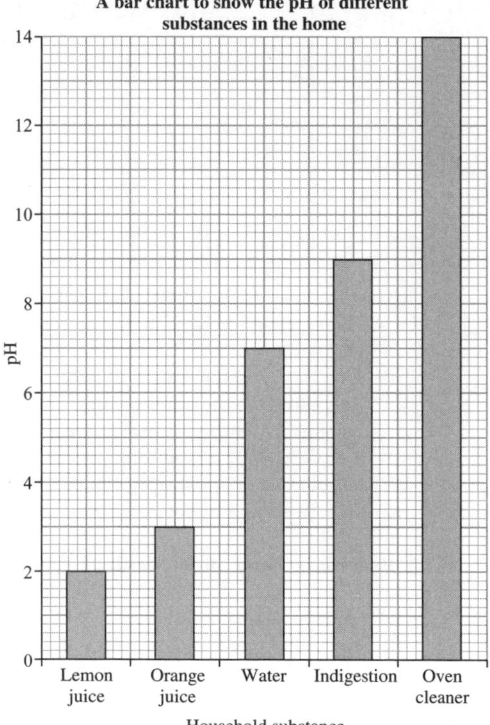

(b) This is a good method of presenting the data as bar charts are usually used when the independent variable is qualitative and the dependent variable is quantitative.

31. Bar charts 2

1 There is no title. This should be included to provide a summary of the chart content.
The axes are not labelled so it is not clear what the data are representing.
There are no units on the y axis so it is not possible to tell how long it takes for the analgesics to relieve headaches.

2

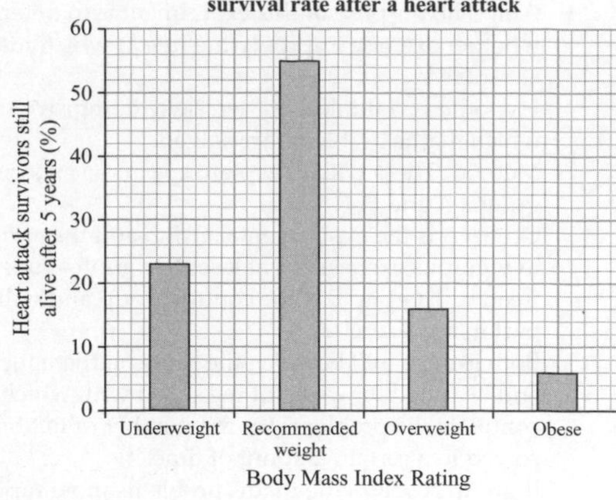

32. Pie charts

1 B

2 To construct a pie chart you need to calculate the angles for the different sectors of the circle, which have to add up to 360°.
Work out the total number in the sample.
Divide the number in each category by the total number in the sample and multiply by 360.
Use a compass to draw the circle and a protractor to measure the angles for each sector.

3 $65 + 10 + 25 = 100$
$65/100 \times 360° = 234°$
$10/100 \times 360° = 36°$
$25/100 \times 360° = 90°$

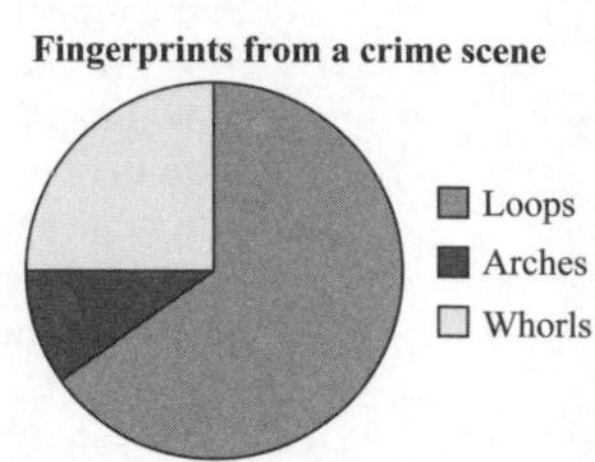

4 The chart needs a title to state what data it is displaying.
Each sector should be labelled or there should be a key so that it is clear what data each sector represents.

33. Line graphs 1

1

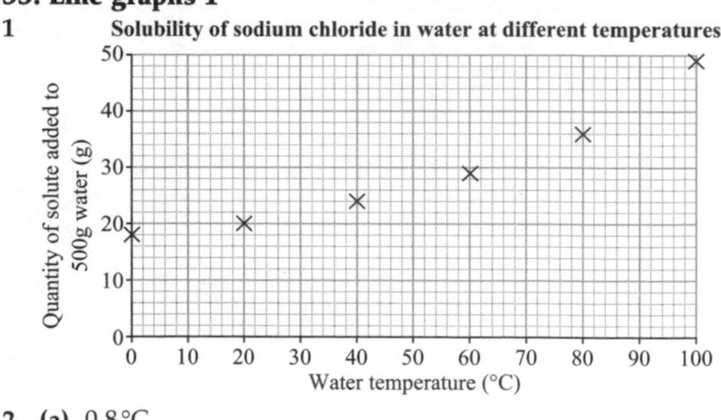

2 (a) 0.8 °C.
(b) 2 hours

34. Line graphs 2

1 B

2 (a)

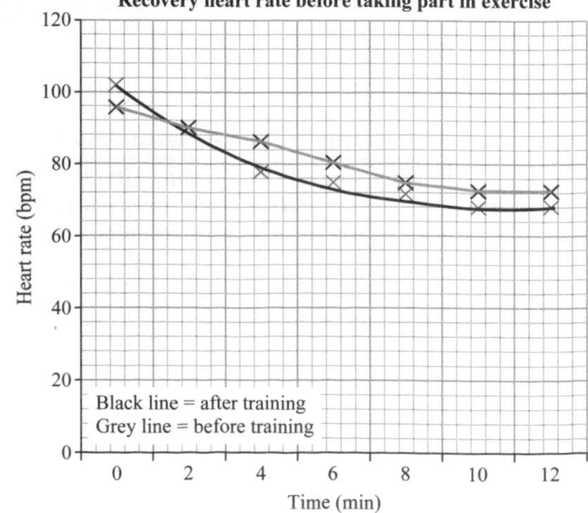

(b) Both heart rates decrease after having stopped exercising. After the training programme, the heart rate reduces to a greater extent in a shorter period of time and returns to resting levels at a faster rate in comparison to before the training programme.

35. Straight lines of best fit

1 D

2

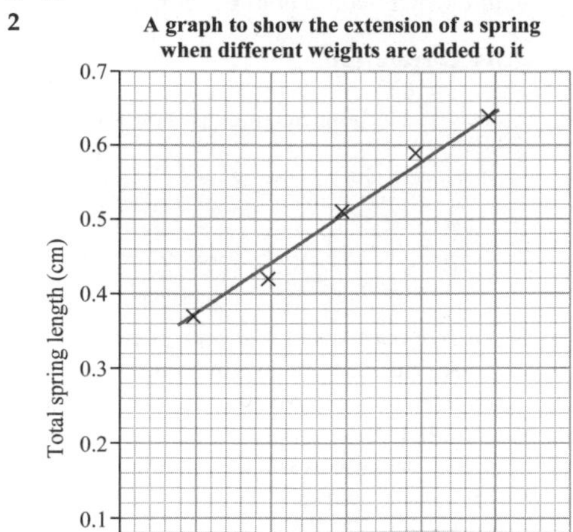

36. Curves of best fit

1 B

2 When very few points on a graph seem to lie close to a straight line, the best way to show the pattern of results is to draw a curve of best fit.

3 (a) and (b)

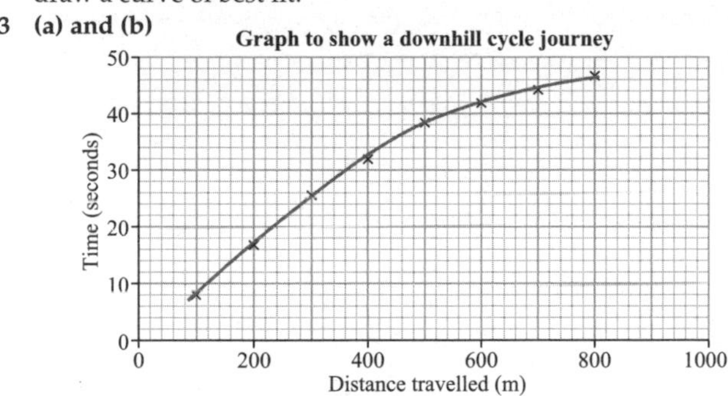

37. Selecting types of graph 1

1 C

2 A bar chart would be best. This is because a bar chart is usually used when the independent variable is qualitative and the dependent variable is quantitative. In this investigation, the independent variable is the different household substances and these are all qualitative and the pH is quantitative.

3 (a)

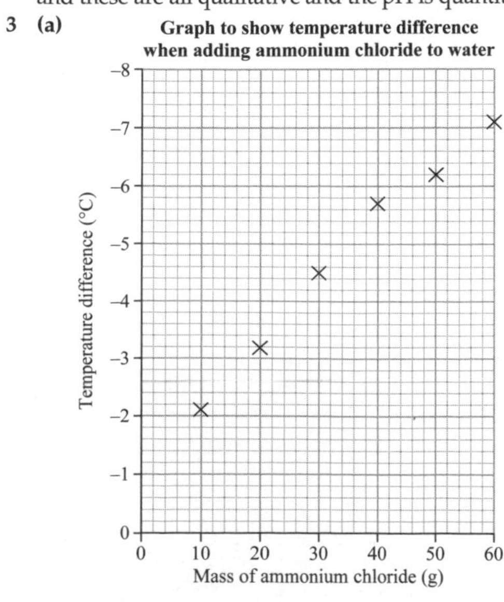

(b) Both sets of variables are quantitative data so a line graph is best. As the dependent variable is decreasing in line with the independent variable, a scatter plot with a line of best fit would be the best way of displaying this trend in the data as this shows how adding more ammonium chloride increases the temperature difference of the water so a clear trend is shown.

38. Selecting types of graph 2

1

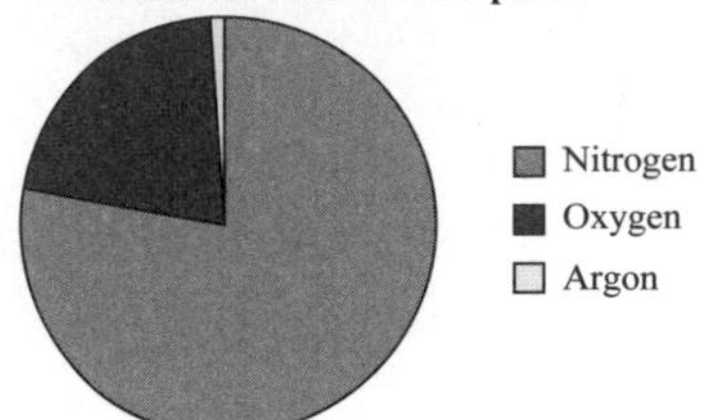

2 (a)

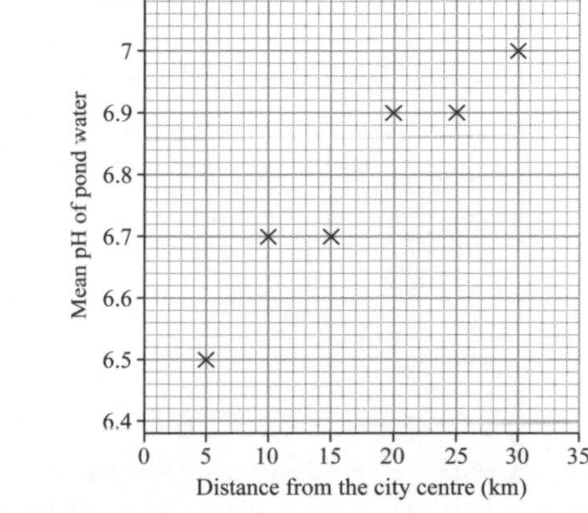

(b) The further away from the city centre, the higher the pH of the pond water/less acidic the pond water.

(c) Yes, a scatter plot with a line of best fit would be the best way of presenting this trend in the data to show that the further away from the city centre the higher the pH of the pond water.

39. Finding values from graphs 1

1 (a) and (b)

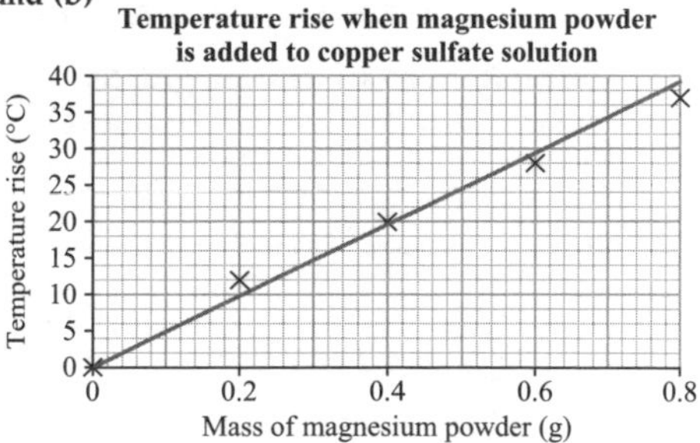

(c) (i) 15°C

(ii) 34°C

40. Finding values from graphs 2

1 A

2 (a) 105

(b) 122

(c) 111

(d) You can only estimate these values as the extrapolation of the line is based on the trend shown in the data that has been gathered. This trend may not continue in line with additional values outside of the data that has been gathered.

41. Calculating gradients from graphs

1 D

2 Change in y-axis values = 16 – 2 = 14
Change in x-axis values = 1.6 – 0.2 = 1.4
Gradient = change in y-axis value/change in x-axis value.
= 14/1.4 = 10 ohms

3 (a) Speed of a two-wheeled electronic scooter

(b)

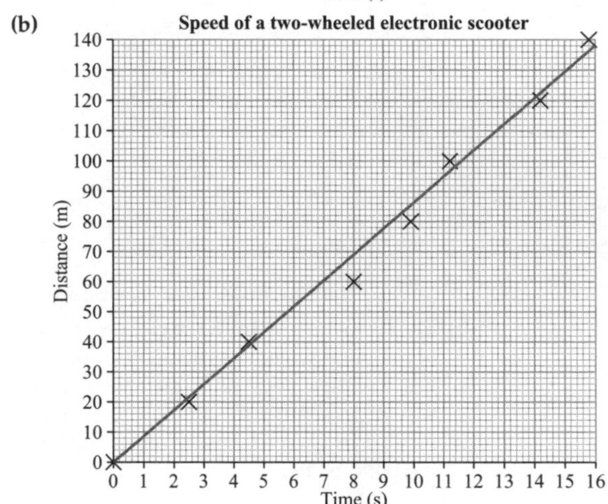

(c) Speed = 40/4.56
= 8.77 m/s (depends on line of best fit drawn).

42. Calculation of areas from graphs

1 (a) (30 – 10) × 20 = 20 × 20 = 400 metres
(b) 0.5 × 10 × 20 = 100 metres
(c) 0.5 × (70 – 30) × 20 = 400 metres

2 (a)

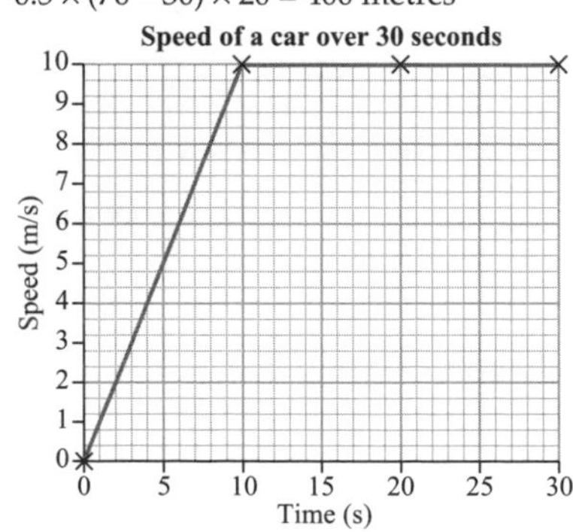

(b) 10 × 10 = 100 m
(c) 10 × 10/2 = 5 m

43. Why anomalous results occur

1 C

2 (a) Person 5 test results with no caffeine = 0.35 seconds.
(b) It is anomalous because it does not fit the pattern of the rest of the results and is out of the range of the other results. All the other results show that reaction time increases after having the caffeinated drink, but this value is faster than the values for having the caffeinated drink. It is also much faster than all of the results for this person so it is outside the range of the results for this person's data.
(c) Any two from the following:
- measurements taken incorrectly
- not following the method and quantities correctly
- not controlling some of the variables
- incorrect recording of results.

(d) This value should be excluded from these calculations because it will have a significant effect on the mean value, which would not be accurate.

44. Positive correlation

1 B

2 (a)

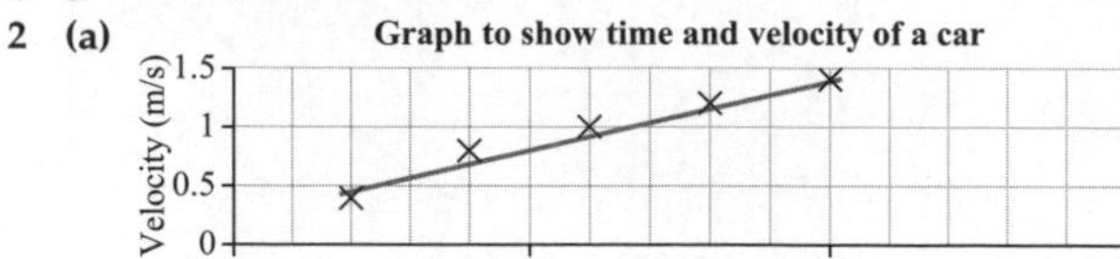

(b) A positive correlation is shown because when the time increases the velocity also increases.

3 (a) The variables are correlated as there is a clear pattern on the graph. It shows a positive correlation between the variables because as the independent variable increases, the dependent variable increases so the line slopes upwards.
(b) The more sugar a person eats the more tooth decay they will experience.

45. Negative correlation

1 C

2 B

3 (a)

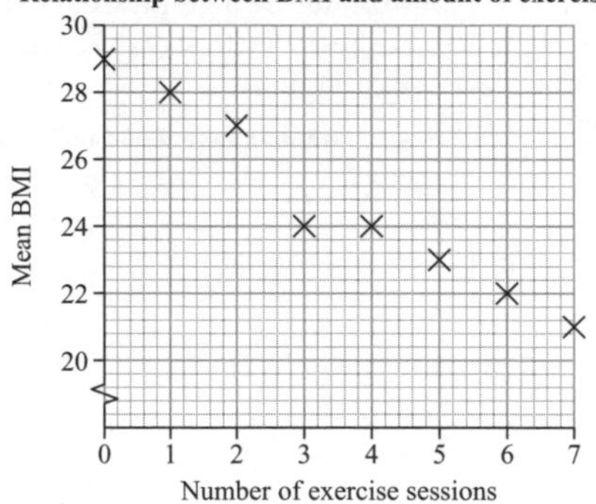

(b) There is a negative correlation because as the amount of exercise increases, the BMI decreases.

46. Direct proportion

1 D

2 (a)

Mass (g)	Length of spring (mm)			Mean length (mm)	Change in length (mm)
	Trial 1	Trial 2	Trial 3		
0	10	10	10	10	0
5	14	15	14	14	4
10	21	22	22	22	12
15	33	34	34	34	24
20	50	49	49	49	39
25	71	70	70	70	60

(b) (i)
(ii)

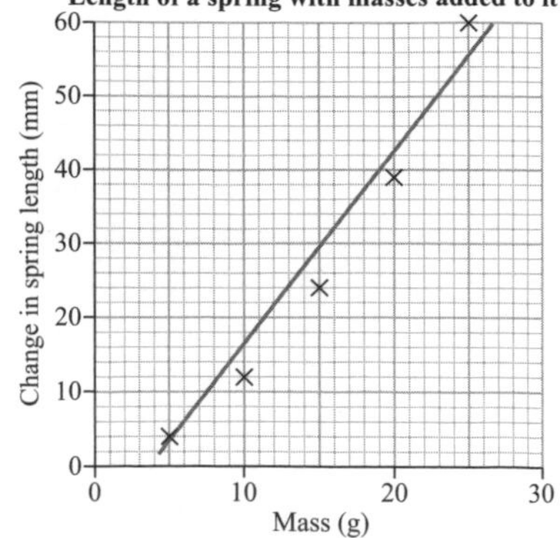

(c) The two variables are directly proportional because when one changes the other changes in the same way, by the same proportion. The line of best fit also passes through the origin of the graph which means the variables are directly proportional.

47. Inverse proportion

1 C

2 **(a)** This is showing an inversely proportional relationship.

(b) As the concentration of solution B increases, the time for the reaction to go to completion decreases.

48. Using evidence 1

1 A

2 Species A lost 0.02 g of water, species B lost 0.01 g of water, species C lost 0.00 g of water.
Therefore species A lost the most water, so species A is the least efficient at retaining water.

3 **(a)** Tyre B has the greatest durability as it has the deepest tread depth after driving around the test track 10 times. It has only lost 0.06 mm compared to the other tyres, which lose between 0.16 mm and 0.52 mm.

(b) No, the test that was conducted was investigating the tread depth of the car tyres, which was exploring the durability of the car tyres. The grip of the tyres was not tested in this investigation so this conclusion cannot be drawn.

49. Using evidence 2

1 Molecules with longer hydrocarbon chains have higher boiling points. Bitumen is the hydrocarbon with the longest hydrocarbon chain as it has the highest boiling point, methane gas has the shortest hydrocarbon chain as it has the lowest boiling point. This is because the longer chained hydrocarbons have stronger forces between the molecules so more energy is needed to separate them.

2 No, the test used is a test for an alkene substance and it has tested positive so the substance is an alkene not an alcohol.

50. Learning aim B sample questions

1 **(a)**

Air pressure (atm)	Maximum Height (cm)			Mean height (cm)
1.40	**Trial 1**	**Trial 2**	**Trial 3**	
1.60	120	122	123	122
1.80	142	143	140	142
2.00	162	163	161	162
2.20	182	134	184	183
1.40	202	200	204	202

(b) There is a positive correlation between the two variables as the line slopes upwards. This shows that as the independent variable (the pressure) increases, the dependent variable (the maximum height) also increases.

Learning aim C

51. Writing a conclusion 1

1 **(a)** Accepted

(b) The results show a continued trend with increasing masses of ammonium chloride up to 60 g. The temperature difference increases as more ammonium chloride is added.

(c) When ammonium chloride is added to water there is an endothermic reaction which takes in energy and results in a temperature decrease of the water. The reason that it is an endothermic reaction is because bonds are being made between the ammonium chloride and water which requires energy and results in the decrease in temperature.

52. Writing a conclusion 2

1 The hypothesis was: 'The higher the pressure in the plastic bottle rocket, the greater the height it will reach'. This hypothesis can be accepted. The results show that for every increase in pressure the plastic bottle 'rocket' travelled to a greater height. The graph shows that there is a positive correlation between the variables – as the pressure increased, the height the 'rocket' reached also increased.

This increase in height is a result of the increase in air pressure, which compressed the air inside the bottle and pushed the water out with greater force. The increase in air pressure pushed on the water in the bottle. According to Newton's third law, the water pushed back on the air and the bottle, which means that the water and the bottle moved in opposite directions.

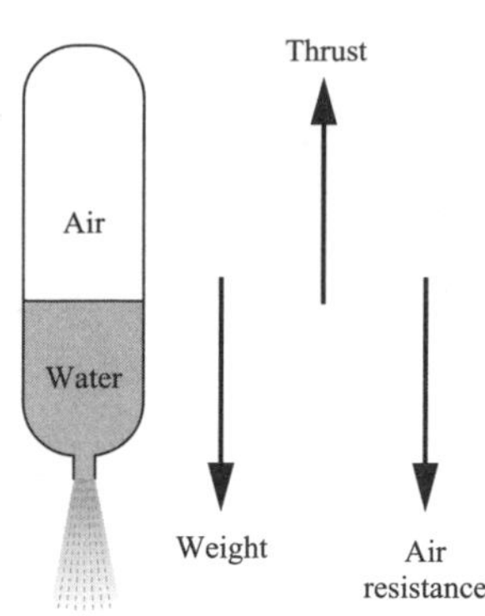

The greater the air pressure in the plastic bottle 'rocket', the higher the rocket will travel.

53. Inferences

1 D

2 **(a)** The more penicillin that the paper disc contains, the larger the area that will be free from bacterial growth.

(b) The more penicillin that is used, the more bacteria will be killed in a prepared agar plate.

(c) A further test would need to be carried out using different quantities of penicillin in different agar plates to see if a stronger does of penicillin kills more bacteria.

54. Support for a hypothesis 1

1 C

2 **(a)** No, the data do not support this hypothesis as the group that took the greatest quantity of ColdK3 had only 2 out of 5000 less chance of catching a cold compared to the people who did not take the ColdK3 drug. Also, the group that took 1 g ColdK3 had the same percentage of people catch a cold as the group that took none of the ColdK3 drug.

(b) D

(c) Group 1.

55. Support for a hypothesis 2

1 **(a)**

Water flea heart rate at different temperatures

Mean water flea heart rate (bpm): 0, 50, 100, 150, 200, 250, 300

Water temperature (°C): 0, 5, 10, 15, 20, 25

(b) There was a positive correlation between the water temperature and the mean heart rate of a water flea. As the temperature of the water rises the heart rate of the water flea increases. It must be a directly proportional relationship as a straight line of best fit can be drawn through the points. But it is not the same as the directly proportional correlation stated in the hypothesis.

(c) As the temperature of the water increases, the mean heart rate of a water flea increases.

56. Support for a hypothesis 3

1 (a)

Pond distance from city centre (km)	Sample pH reading					Mean pH
	1	2	3	4	5	
0	6.5	6.4	6.5	6.5	6.4	6.5
5	6.5	6.5	6.4	6.5	6.5	6.5
10	6.7	6.6	6.5	6.6	6.4	6.6
15	6.8	6.7	6.6	6.8	6.8	6.7
20	6.8	6.8	6.8	6.8	6.7	6.8
25	6.9	6.9	6.8	6.8	6.9	6.9
30	6.9	6.9	6.8	6.7	6.9	6.8

(b)

pH of pond water away from the city centre

Mean pH of pond

Pond distance from city centre (km)

(c) The evidence supports this hypothesis up to a point as the pH of the pond generally increases as the ponds get further away from the city centre.
However, the pH value of the pond decreases slightly at 30 km when compared with 25 km distance from the city centre which means that this hypothesis is not fully supported by this data.

57. Strength of evidence

1 C

2 (a)

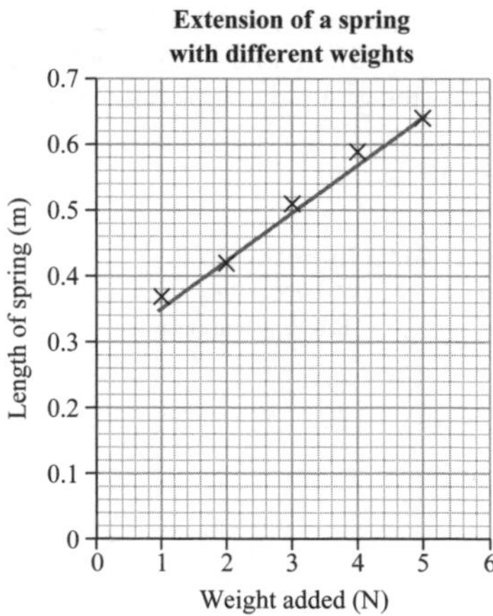

(b) There is a strong positive correlation showing that as more force is added to the spring, the length of the spring increases.

(c) Yes, this hypothesis is strongly supported by this evidence as the graph shows a strong positive correlation between the variables as most of the plotted points are on the line of best fit, or if they are not on the line they are very close to it.

58. Sufficient data 1

1 B

2 (a) The results suggest that second-hand smoke does have a negative effect on the growth rate of a baby in the womb as the mean mass at birth for a baby with a father who smokes is less than the mean mass at birth for a baby with parents who do not smoke.
However, the masses at birth are very close together – there is only 0.2 kg difference between the two values. This means that in some cases the mass at birth of a baby born to non-smokers could have been less than a baby born to parents where the father smokes.

(b) The investigation has only considered if the father smoked, the father may not have smoked in the house or anywhere near the mother whilst she was pregnant so the baby may not have been exposed to second hand smoke which means the data collected is not reliable.

59. Sufficient data 2

1 D

2 A

3 This conclusion cannot be drawn from these data as there are only three data points on the graph, which is not enough for a line of best fit to be drawn or relationship to be stated. If any of the measurements were taken incorrectly then, with only three data points, anomalous data is very difficult to spot.

60. Validity of a conclusion

1 (a) Taking part in 1 minute of skipping will increase a person's heart rate.

(b) A

2 This is not a valid conclusion because the investigation did not measure the rate of photosynthesis of the plants. It measured the height of the plants which may not be directly related to the rate of photosynthesis of the plant.

61. Evaluating an investigation 1

1 (a) No, there are not enough readings as they were only taken every 2 hours. The UV rating increases from 10 am every day but as readings were only taken every 2 hours it is not possible to tell if the UV rating could have been higher at 11 am, 1 pm or 3 pm each day.

(b) If it had been a cloudy day the UV ratings would have been lower because the clouds block out some of the UV rays.

(c) No, the readings were taken during 1 month and only over a 4 day period. This sample is far too small to represent the UV levels in a country as the UV levels may well change at different times of the year during the different seasons. Also, the UV readings were only taken in one place and different parts of the country may receive differing levels of UV rays at different parts of the day depending upon their location in the UK.

62. Evaluating an investigation 2

1 (a) and (b)

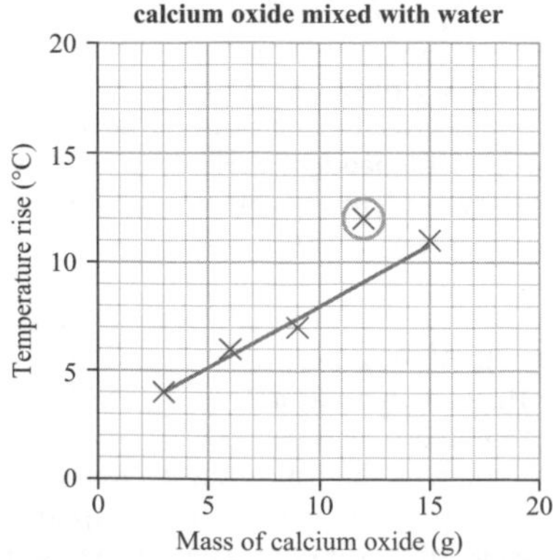

(c) The data are not reliable as each measurement has only been taken once. An investigation should include repeating measurements at least three times to provide reliable data.

(d) Only one reading for each mass of calcium oxide was taken. It would have been better to take a few readings and calculate the mean temperature change which would give more accurate data.

63. Improving an investigation 1

1 (a)

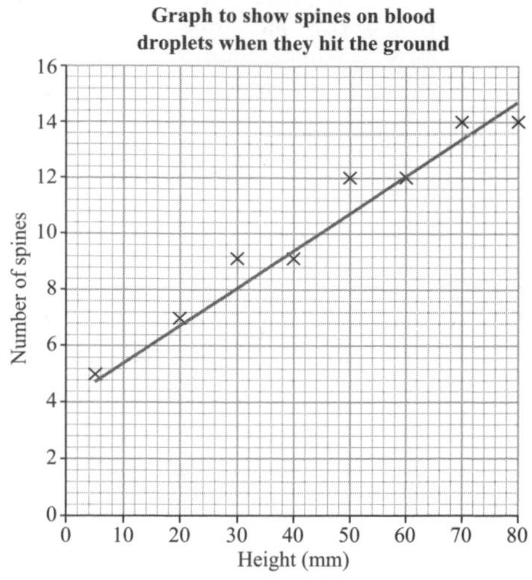

(b) They could increase the range of the independent variable as the heights used are only from 10 mm to 80 mm, which is not very high off the ground. The reason for carrying out this investigation was to find out about the height that a person was at when they started to lose blood so the independent variable should really extend to at least the height of an average person (at least 1 metre).

64. Improving an investigation 2

1 (a) Subjective means that the results depend on a person's opinion and are not actually measured results so they may not be accurate.

(b) The mascara is designed to increase the length of a person's eyelashes. Therefore, instead of asking for a person's opinion on whether they think their eyelashes are longer, they could measure the length of a selected number of eyelashes before and after applying mascara A and mascara B and the results could be compared to see which one does actually increase the length of the lashes.

(c) The larger the sample size in an investigation, the greater the chance of obtaining results that will help to prove or disprove the hypothesis because statistically the larger the sample size, the more significant the results of the experiment are likely to be.

65. Extending an investigation

1 A

2 (a) The percentage of glucose in the sports drink could be changed to see if a higher concentration of glucose will allow a person to exercise for even longer.

(b) The temperature could be changed to see if the sports drink is effective in improving exercise performance in hotter conditions.

(c) The sample is only using males, the sample could be changed to test females to see if females gain the same exercise performance benefits as males from drinking a 9% glucose sports drink.

66. Learning aim C sample questions

1 (a) She should use the same volume of acid each time by measuring it using a measuring cylinder.
She should use the same concentration of acid each time. She can do this by using acid from the same bottle.
She should leave each indigestion remedy in the acid for the same length of time before taking pH reading. She can use a stopwatch to check this time.

(b) The acid used in this investigation is sulfuric acid, which is not the acid that is produced in the stomach that causes indigestion. To improve this experiment it would be better to use hydrochloric acid at the same concentration as would be found in a person's stomach so that the indigestion tablet can be tested on this to see if it really does reduce the acidity.

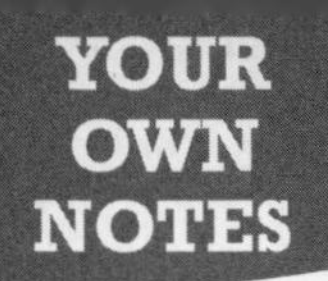

Your own notes

Published by Pearson Education Limited, Edinburgh Gate, Harlow, Essex, CM20 2JE.

www.pearsonschoolsandfecolleges.co.uk

Copies of official specifications for all Edexcel qualifications may be found on the Edexcel website: www.edexcel.com

Edited, produced and typeset by Wearset Ltd, Boldon, Tyne and Wear

Illustrated by KJA Artists
Cover illustration by Miriam Sturdee

First published 2013

17 16 15 14 13
10 9 8 7 6 5 4 3 2

British Library Cataloguing in Publication Data
A catalogue record for this book is available from the British Library

ISBN 978 1 446 90284 4

Printed in Slovakia by Neografia

Acknowledgements
The publisher would like to thank the following for their kind permission to reproduce their photographs:

(Key: b-bottom; c-centre; l-left; r-right; t-top)

Alamy Images: Ashley Cooper 62, blickwinkel 55, Harald Richter 48, MetaPics 64, Stockbyte Platinum 17; **Getty Images:** Image Source 5, Tobias Titz 11; **Science Photo Library Ltd:** Martyn F. Chillmaid 13, Mauro Fermariello 56; **Veer/Corbis:** csjani 29, Shaheed 21br, Skystorm 28, warrengoldswain 21bl

In some instances we have been unable to trace the owners of copyright material, and we would appreciate any information that would enable us to do so.

In the writing of this book, no Edexcel examiners authored sections relevant to examination papers for which they have responsibility.